AF247579

Reviews of Physiology Biochemistry and Pharmacology

124

Editors

M.P. Blaustein, Baltimore • H. Grunicke, Innsbruck
E. Habermann, Gießen • D. Pette, Konstanz • H. Reuter, Bern
B. Sakmann, Heidelberg • M. Schweiger, Berlin
E.R. Weibel, Bern • E.M. Wright, Los Angeles

With 6 Figures and 7 Tables

Springer-Verlag
Berlin Heidelberg New York London Paris
Tokyo Hong Kong Barcelona Budapest

ISBN 3-540-57587-1 Springer-Verlag Berlin Heidelberg New York
ISBN 0-387-57587-1 Springer-Verlag New York Berlin Heidelberg

Library of Congress-Catalog-Card-Number 74-3674

Typesetting: Data conversion by Springer-Verlag

27/3130-5 4 3 2 1 0 – Printed on acid-free paper

Contents

Indexed in Current Contents

Rev. Phyiol. Biochem. Pharmacol., Vol. 124
© Springer-Verlag 1994

Complexity and Versatility of the Transcriptional Response to cAMP

V. DELMAS, C. A. MOLINA, E. LALLI, R. DE GROOT,
N. S. FOULKES, D. MASQUILIER, and P. SASSONE-CORSI

Contents

Laboratoire de Génétique Moléculaire des Eucaryotes, CNRS, U184 INSERM, Institut de Chimie Biologique, Faculté de Médécine 11, rue Humann, 67085 Strasbourg, France

1 Introduction

The regulation of gene expression by specific signal transduction pathways is closely connected to the cell phenotype. The response elicited by a given transduction pathway varies according to the cell type. The finding that most of the known nuclear oncogenes encode proteins involved in the regulation of gene expression inspired the concept that the aberrant expression of some key genes could cause cellular transformation or altered proliferation (Lewin 1991). The study, and ultimately the understanding, of these processes will hopefully help us to unravel the profound changes that cause cancer and by the same token the physiology of normal growth.

An important step toward the comprehension of how the function of transcription factors can be modulated has been the discovery that many constitute final targets of specific signal transduction pathways, activated intracellularly by various signals at the cell surface. The two major signal transduction systems are those including cAMP and diacylglycerol (DAG) as secondary messengers (Nishizuka 1986). Each pathway is also characterized by specific protein kinases [protein kinase A (PKA) and protein kinase C (PKC), respectively] and its ultimate target DNA control element, the cAMP-responsive element (CRE) and the TPA-responsive element (TRE). Although initially characterized as distinct systems, accumulating evidence points toward extensive cross-talk between these two pathways (Cambier et al. 1987; Yoshimasa et al. 1987). Here we focus primarily on the targets of the cAMP-mediated transduction response.

2 The PKA Signal Transduction Pathway

Intracellular levels of cAMP are regulated primarily by adenylyl cyclase. This enzyme is in turn modulated by various extracellular stimuli mediated by receptors and their interaction with G proteins (McKnight et al. 1988). The binding of a specific ligand to a receptor results in the activation or inhibition of the cAMP-dependent pathway, ultimately affecting the transcriptional regulation of various genes through distinct promoter-responsive sites (Montmayeur and Borrelli 1991). Increased cAMP levels directly affect the function of the tetrameric PKA complex (Krebs and Beavo 1979). Binding of cAMP to two PKA-regulatory subunits releases the catalytic subunits enabling them to phosphorylate target proteins. A number of isoforms for both the regulatory and catalytic subunits have been identified, suggesting a further level of complexity in this response (McKnight et al. 1988). In the nucleus the phosphorylation state of transcription factors and related proteins appears directly to modulate their function and thus the expression of cAMP-inducible genes.

3 Promoter Sites of cAMP-Inducible Genes

3.1 The cAMP-Responsive Element

A promoter element that mediates the response to increased levels of intra-cellular cAMP is the CRE (Comb et al. 1986; Andrisani et al. 1987; Dele-geane et al. 1987; Sassone-Corsi 1988). A consensus CRE site constitutes an 8-bp palindromic sequence (TGACGTCA). Several genes which are regu-lated by a variety of endocrinological stimuli contain similar sequences in their promoter regions although at different positions. A comparison of the CRE sequences identified to date shows that the 5' half of the palindrome, TGACG, is the best conserved, whereas the 3' TCA motif is less constant. The binding site specificity appears to require 18–20 bp, since the five or so bases flanking the core consensus have been shown to dictate, in some cases, the permissivity of transcriptional activation (Deutsch et al. 1988). In many genes the CRE sequence is located in the first 200 bp upstream from the cap site. In most cases there is only one CRE element per promoter, although there are notable exceptions. The promoter of the α-chorionic gonadotropin gene, for instance, contains two identical, canonical CREs in tandem, be-tween positions –117 and –142 (Delegeane et al. 1987). The promoter of the pituitary-specific transcription factor GHF-1/Pit-1, on the other hand, con-tains two different CREs between positions –200 and –150, which are separated by a 40-bp spacer (McCormick et al. 1990). The proto-oncogene c-*fos* contains a powerful CRE at position –60 (Sassone-Corsi et al. 1988a), but other CRE-like sequences are also present within the gene regulatory region (Berkowitz et al. 1989); however their precise function has yet to be determined.

The CRE consensus sequence also appears in the context of other pro-moter elements. These include the ATF sites in the early promoters of adenovirus (Lin and Green 1988; Sassone-Corsi 1988), the 21-bp tax-de-pendent enhancer of the human T-cell leukemia virus (HTLV-I) virus long terminal repeat (LTR; Yoshimura et al. 1990), and the X-box motif associ-ated with MHC class II genes (Liou et al. 1988).

3.2 Non-CRE Sites

Recently an additional sequence responsive to increased cAMP levels has been described in the promoter of the steroid hydroxylase gene CYP17 (Zanger et al. 1991). Although no information is available yet on the regula-tory proteins which bind to this novel element, it is tempting to speculate that more than one class of sequence element can mediate the response to cAMP, possibly in a development- or cell-specific manner. In further support of this notion, another non-CRE element which also mediates cAMP induction is the AP-2 recognition site (Imagawa et al. 1987).

4 Structure of the Nuclear Effectors of the PKA Pathway

An important step toward the understanding of cAMP-regulated gene transcription has been made with the cloning of cDNAs encoding CRE binding proteins, thereby allowing studies on the precise structure-function relationship of these factors. At least ten CRE binding factor cDNAs have been isolated. They were obtained by screening a variety of cDNA expression libraries with CRE and ATF sites, the HTLV-I LTR 21-bp repeat, and the MHC class II X-box sequences. They all belong to the basic region/leucine-zipper (b-Zip) family of proteins. It has been demonstrated that, as in the case of Fos, Jun and C/EBP, the leucine zipper is responsible for the dimerization of the protein, and that dimerization is a prerequisite for DNA binding. The model by Vinson et al. (1989) suggests the presence of a *bipartite* DNA binding domain, as dimerization ensures the correct orientation of the adjacent basic regions in order to allow their optimal contact with the recognition sequence. It has been defined that the basic region, 50% rich in lysine and arginine residues, is infact divided into two subdomains containing clusters of basic residues separated by a "spacer" of alanines, conserved among all leucine zipper transcription factors. In this model these two regions recognize the two halves of the palindromic recognition sequence. The positively charged amino acids in the basic region lie on one face of the two helices of a helix-bend-helix structure. The two positively charged α-helices lie in the major groove of the DNA helix positioned so that the positive charges are in contact with the negative charges of the phosphate backbone (Vinson et al. 1989).

All these different proteins are highly homologous in their b-Zip region, while they diverge in the other parts of the proteins. Based on common regions of homology, these factors can be divided into subfamilies. For example, CREB, ATF1, and CREM share extensive regions of homology (Foulkes et al. 1991a; Hai et al. 1989) but are clearly distinct from ATFa and CRE-BP1, which constitute another subfamily (Ivashkiv et al. 1990; Gaire et al. 1990).

The b-Zip is responsible for DNA binding as well as for dimerization of the proteins. The different factors are able to heterodimerize with each other but only in certain combinations. A "dimerization code" exists that seems to be a property of the leucine zipper structure of each factor. For example, ATF1 can dimerize with CREB but not with CREBP1 or ATF3, and

CREBP1 can heterodimerize with ATF3 but not with CREB (Hurst et al. 1991; Hai et al. 1989; Hoeffler et al. 1991). Although most of these heterodimers have not yet been identified in vivo, these observations suggest that part of the complexity of CRE-dependent regulation is accounted for by the large number of homo- and heterodimers combinations possible.

5 Model of Transcriptional Activation

5.1 The CREB Gene

The identification and cloning of the CRE binding protein (CREB) gene opened the way for a molecular analysis of the trans-activation phenomenom. The CREB cDNAs have been cloned from human placenta and rat brain libraries (Hoeffler et al. 1988; Gonzales et al. 1989). CREB is ubiquitously expressed, suggesting a housekeeping role for this factor.

5.2 Activation by Phosphorylation

An important insight into the molecular mechanisms by which the transcription of CRE-containing genes is induced, came from experiments demonstrating that upon activation of the adenylyl cyclase pathway, a serine residue at position 133 of CREB is phosphorylated by PKA (Fig. 1; Gonzales and Montminy 1989). The phosphorylation appears indispensable for activation, and the phosphoserine cannot be substituted by other negatively charged residues (Yamamoto et al. 1988; Gonzales and Montminy 1989; Lee et al. 1990). Whether phosphorylation by PKA modulates DNA binding by CREB is a slightly controversial point. Indeed, Yamamoto et al. (1988) indicate that PKA-mediated phosphorylation of CREB does not affect DNA binding. By contrast, Nichols et al. (1992) have reported that phosphorylation of CREB by PKA causes a modest increase in binding to high-affinity CRE sites but a stronger enhancement in binding to low-affinity CREs. However, the major effect of phosphorylation rather seems to occur at the level of the trans-activation function of CREB. It has been proposed that this could happen by inducing a conformational change of the protein (Gonzalez et al. 1991). However, in contrast to this hypothesis, recent studies by Leonard et al. (1992) have shown that CREB can be a very potent activator in the absence of phosphorylation in the pancreatic islet cell line Tu6. The mechanism of this phosphorylation-independent activity remains to be determined. Interestingly, alternative signal transduction pathways can also induce phosphorylation of serine 133. In PC12 cells, increases in the levels

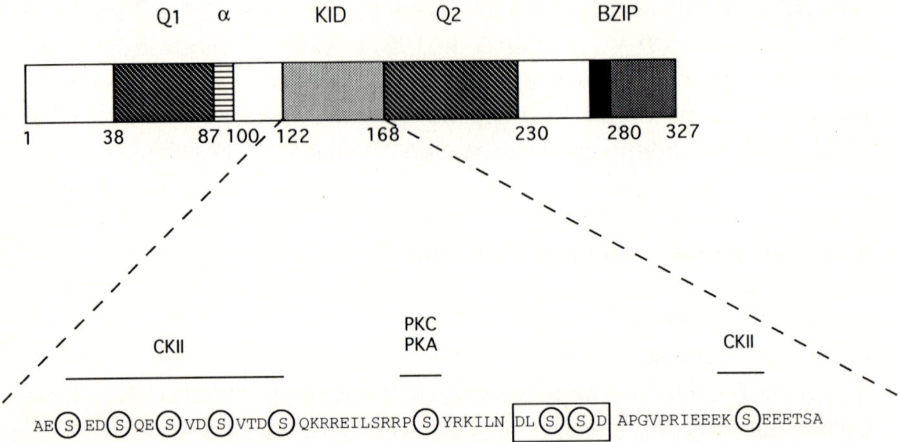

Fig. 1. Structure of transcriptional activator CREB and localization of the phosphorylated residues. The detailed structure of the kinase-inducible domain (*KID*) and the serine residues (*circles*) phosphorylated by the indicated kinases are shown: *PKA*, protein kinase A; *PKC*, protein kinase C; *CKII*, casein kinase II. *Shaded boxes*, KID , α-region (α), and flanking glutamine-rich domains (*Q*). *BZIP*, The basic domain and leucine zipper of the DNA binding region. Numbers of amino acids are given beneath each domain

of intracellular Ca^{2+} by membrane depolarization cause phosphorylation of serine 133 and a concomitant induction of c-*fos* gene expression mediated by a CRE in the c-*fos* promoter (Sassone-Corsi et al. 1988a; Sheng et al. 1990, 1991). CREB mutants lacking serine 133 were unable to activate c-*fos* transcription (Sheng et al. 1991). Although Ca^{2+} calmodulin-dependent (CAM) kinases were shown to be able to phosphorylate serine 133 in vitro (Sheng et al. 1991; Dash et al. 1991), their role in vivo remains unclear, since PKA seems to be necessary for c-*fos* induction by Ca^{2+} influx in PC12 cells (Ginty et al. 1991). In addition, CREB is also phosphorylated upon stimulation by TGFβ1, although the target residue remains to be determined (Kramer et al. 1991).

Recent experiments by Hagiwara et al. (1992) propose a mechanism to explain the attenuation of CREB activity following induction by forskolin. Their results indicate that after the initial burst of phosphorylation in response to cAMP, CREB is dephosphorylated in vivo by protein phosphatase-1 (PP-1), and transcription of the somatostatin gene is correspondingly reduced. However, Nichols et al. (1992) show that both PP-1 and PP-2A can dephosphorylate CREB in vitro, resulting in a decreased binding to low-affinity CRE sites in vitro. Therefore, the precise role of PP-2A in the dephosphorylation of CREB remains to be determined.

The structure of the transcriptional activation domain of CREB includes more than the phosphoacceptor region (see Fig. 1). Serine 133 is located in a region of about 50 amino acids containing an abundance of phosphorylated serines and acidic residues, the phosphorylation box (P-Box) or kinase inducible domain (KID), which was shown to be essential for *trans*-activation by CREB (Lee et al. 1990; Gonzalez et al. 1991). Although phosphorylation of serine 133 appears indispensable for activation by CREB, it is not sufficient for full activity. An acidic region just downstream of serine 133 (140-DLSSD) was shown to be important for CREB function (Lee et al. 1990; Gonzalez et al. 1991). In addition, deletion of a region called $\alpha 2$, containing several sites that can be phosphorylated by CKII in vitro, caused a decrease in CREB activity, although differences in the magnitude of this decrease were reported (Lee et al. 1990; Gonzalez et al. 1991).

Interestingly, two other CRE binding factors have been reported to be activated by PKA. ATF1 was shown to activate transcription after cotransfection of the catalytic subunit of PKA (Rehfuss et al. 1991; Flint et al. 1991). Although ATF1 can be phosphorylated by PKA in vitro, in vivo phosphorylation has yet to be demonstrated. The recently described activator isoform of CREM, CREMτ, can also mediate cAMP-induced transcription (Foulkes et al. 1992; see below). CREMτ is phosphorylated by PKA in vitro as well as in vivo on serine 117, the counterpart of serine 133 in CREB (deGroot et al. 1993).

5.3 The Roles of the Glutamine-Rich Domains

Flanking the P-box are two regions in which there are about three-times more glutamine residues than in the remainder of the protein (see Fig. 1). Glutamine-rich domains have been characterized in other factors, such as AP-2 and Sp1 (Williams et al. 1988; Courey and Tjian 1989) as transcriptional activation domains. The significance of the first glutamine-rich domain (Q1) is not completely clear, since it was reported to enhance CREB activity by Gonzalez et al. (1991) while Lee et al. (1990) failed to find an effect when they deleted this region. However, this apparent contradiction might be caused by the different CREB isoforms studied by these two groups, since Gonzalez et al. studied CREBα/341, while Lee et al. have used CREBD/327. Interestingly, two different CREM isoforms containing either Q1 (τ1) or Q2 (τ2) are both transcriptional activators (Fig. 2). The Q2 domains appears to confer a slightly higher activation potential than the Q1 domain (Laoide et al. 1993). These results demonstrate that the Q1 and Q2 regions probably function in an additive manner to generate the full activation potential of CREMτ.

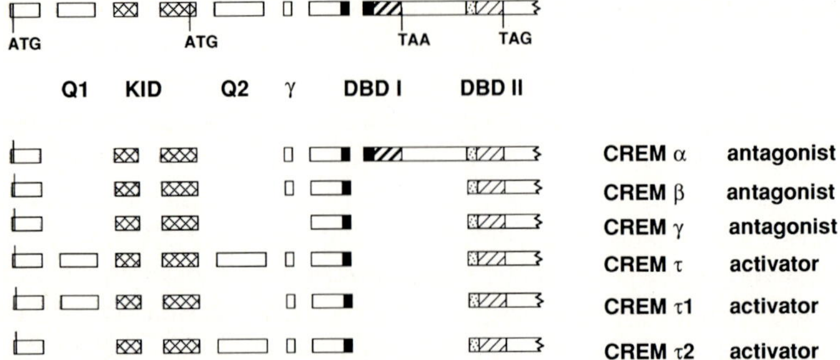

Fig. 2. CREM multiexonic structure generates isoforms with different functions. Schematic representation of CREM exon structure and transcripts. *Top row*, the various exons derermine the functional domains of the proteins (Laoide et al. 1993). *Q1, Q2*, glutamine-rich domains; *KID*, kinase-inducible domain, γ, a 12 amino acid domain lacking in the CREMγ isoform and which has no counterpart in CREB; *DBDI, DBDII*, the two alternative DNA binding domains. *Hatched boxes*, the leucine zipper portion of the two DNA binding domains; *solid black boxes*, the basic region of DBDI and the shared 5' portion of the basic domain of DBDII. The position of the initiation (ATG) and termination codons (TAA, TAG) are indicated. The position of the alternative ATG used to generate S-CREM (Delmas et al. 1992) is also shown. Beneath is also shown the exon composition of the various CREM isoforms

The current model to explain the activation of CREB suggests that upon phosphorylation of serine 133 by PKA, a conformational change is induced, which leads to exposure of the glutamine-rich activation domains (Gonzalez et al. 1991). The other regions which were identified as being important for CREB function might be involved in correctly spacing the phosphorylation site with respect to the glutamine-rich domains. Verification of this model awaits determination of the crystal structure of both unphosphorylated as well as phosphorylated CREB.

6 Molecular Basis of Negative Regulation: The CREM Gene

The discovery of the cAMP-responsive element modulator (CREM) gene opened a new dimension in the study of transcriptional response to cAMP (Foulkes et al. 1991a). This is due to the remarkable genomic structure of the CREM gene, which offers clues to the understanding of the generation of functional diversity in transcription factors. CREM is also the first CRE binding protein with antagonistic function.

6.1 A Remarkable Genomic Structure and Cell-Specific Expression

The CREM gene was isolated from a mouse pituitary cDNA library screened at low stringency with oligonucleotides corresponding to the leucine-zipper and basic region of CREB. The logic behind this approach is that the adenylyl cyclase pathway plays an important role in the modulation of the hormonal regulation in the pituitary gland. The most striking feature about the CREM cDNA is the presence of two DNA binding domains. The first is complete and contains a leucine zipper and basic region very similar to CREB; the second is located in the 3' untranslated region of the gene, out of phase with the main coding region, and contains a half basic region and a leucine zipper more divergent from CREB. Various mRNA isoforms were identified that appear to be obtained by differential cell-specific splicing (Fig. 2). Alternative usage of the two DNA binding domains was demonstrated in various tissues and cell types, where quite different patterns of expression were found (Foulkes et al. 1991a). This strongly contrasts with CREB and ATFs expression which is ubiquitous (Hai et al. 1989; Habener 1990), suggesting their roles as constitutive regulators.

CREM expression appears to be finely regulated, both transcriptionally and posttranscriptionally. In fact, not only cell- and tissue-specific expression is observed, but also the production of various isoforms. Four major isoforms have been characterized, of which three act as repressors and one as a powerful activator. The three antagonistic products were the first to be described (Foulkes et al. 1991a). These isoforms revealed alternative usage of the two DNA binding domains (α and β isoforms), as well as a small deletion of 12 amino acids (γ isoform; Fig. 2). The potential of even more complexity of CREM regulation is hinted at by the possible usage of alternative poly(A) addition sites and by the presence of ten AUUUA sequences in the 3' untranslated region, elements thought to be involved in mRNA instability (see Fig. 3; Shaw and Kamen 1986). The strict cell- and tissue-specific expression of CREM is indicative of a pivotal function in the regulation of cell-specific cAMP response. This suggests that CREM occupies a central control point in the pituitary, since it is known that the physiology of this gland is finely regulated by a multiplicity of hormones whose coupled signal transduction pathways involve adenylate cyclase. Interestingly, other well-described examples of cell-specific splicing include the genes encoding neuronal peptides and hormones in brain and pituitary cells (Leff et al. 1986). It appears clear, thus, that cell-specific splicing is a crucial mechanism of CREM regulation, which modulates the DNA binding specificity and activity of the final CREM products.

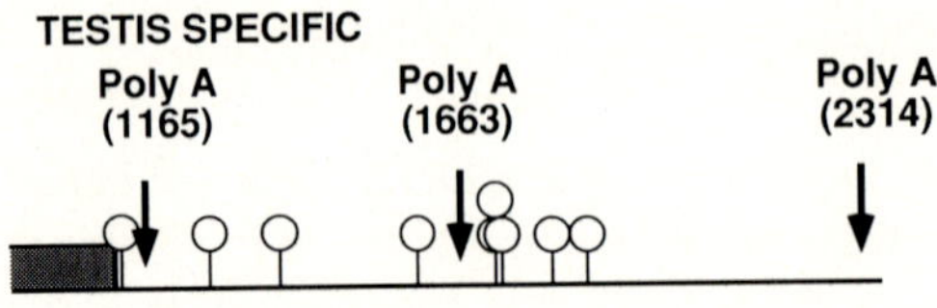

Fig. 3. Schematic representation of the 3' untranslated region of the CREM gene (Foulkes et al. 1991a). Three alternative polyadenylation signals are used (*PolyA*); their position within the CREM cDNA is indicated. The farthest upstream PolyA site is used predominantly in male germ cells. In the region there are ten destabilizer sequences AUUUA (*open circles*; Shaw and Kamen 1986)

6.2 Antagonists and Activators from the Same Gene

The first cDNA clones which were characterized from the CREM gene encode antagonists of cAMP-induced transcription (Foulkes et al. 1991a). The CREM antagonists share extensive homology with CREB, but they lack two glutamine-rich domains, which have been shown to be necessary for transcriptional activation in CREB (Gonzalez et al. 1991). Interestingly, the CREM gene also encodes an activator of transcription (Foulkes et al. 1992). In the adult testis, an isoform CREMτ has been identified which resembles in structure one of the antagonist forms (CREMβ) but includes two exons that encode two glutamine-rich domains. As mentioned before, this form has been demonstrated to transactivate transcription from a CRE site, as well as the CREM isoforms containing either Q1 or Q2 glutamine-rich domains (Laoide et al. 1993).

The CREM proteins specifically recognize CREs and show the same binding properties as CREB. This is not surprising, considering the high homology in the DNA binding domains between these proteins. CREM proteins containing either DNA binding domain I or II heterodimerize with CREB (Foulkes et. al. 1991a; Laoide et al. 1993), although it appears that CREMα-CREB heterodimer formation is more favored than CREMβ-CREB. These notions suggest that CREM proteins might occupy CRE sites as CREM dimers or as CREM-CREB heterodimers, thus generating complexes with altered transcriptional functions. Infact CREM products act by impairing CRE-mediated transcription, and as such are considered as antagonists of cAMP-induced expression. In transfection experiments, using CRE reporter plasmids, it was demonstrated that CREM antagonists block the transcriptional activation obtained by the joint action of CREB or CREMτ and the catalytic subunit of the cAMP-dependent PKA (Mellon et al. 1989). These observations strongly support the notion that CREM proteins negatively modulate CRE promoter elements in vivo. An important question is how CREM proteins work. The two most likely hypotheses are as follows. According to the first scenario, CREM proteins dimerize and bind to CRE sites. Downregulation is achieved by the occupation of these sites, which are unavailable for CREB or CREMτ. Similarly, if CREB is already bound, CREM proteins might squelch them because of their possible higher affinity for a specific site. According to the second model, CREM proteins are able to dimerize with CREB to generate non-functional heterodimers. Negative regulation is achieved by titrating active CREB molecules, and CREM proteins could act as activator traps. Since both CREM dimers and CREB-CREM heterodimers bind to CRE sites, both hypotheses are justified and both mechanisms may operate. The CREM mRNA isoforms are a graphic illustration of how alternative splicing can modulate the function of a transcription factor in a tissue- and developmental-specific manner (Foulkes and Sassone-Corsi 1992).

6.3 Alternative Translation Initiation:
Another Way to Generate a Repressor

We have also shown that by alternative usage of translation initiation sites, a single CREM mRNA generates both an activator and a repressor (Delmas et al. 1992). The use of an internal AUG in the CREMτ transcript generates S-CREM, a protein that acts as a powerful repressor of cAMP-induced transcription. It is puzzling that CREM, which already makes extensive use of differential splicing to generate both an activator and antagonists (Fig. 2), utilizes an additional mechanism to generate factors with opposite functions. It is important to note, however, that S-CREM is distinct from the antagonist CREM forms generated by alternative splicing. It does not contain the phosphoacceptor sites (P-box), but instead contains a single glutamine-rich domain. This is in contrast to the CREMα/β/γ antagonists, which contain the P-box but no glutamine-rich domains (Foulkes et al. 1991a). It is reasonable to hypothesize that the structural differences present between these various downregulators have functional significance, and that these may interact differently with other components of the transcriptional machinery. In addition, the differential presence of the P-box among these various CREM downregulators, could suggest that some of them might be modulated in their function by phosphorylation.

Internal initiation has been described for others genes, for example the oncogenes *int*2 (Acland et al. 1991), *pim*-1(Saris et al. 1991) and the androgen receptor *Tfm* (Gaspar et al. 1991). Interestingly, alternative initiation has been described for another leucine-zipper regulatory protein, the Liver Activator Protein (LAP; Descombes et al. 1991). In LAP, an internal AUG is used to produce LIP, a repressor protein. The major difference between LAP and CREM is that LAP is an intronless gene, whereas CREM has a multiexonic structure (Laoide et al. 1993; and in preparation), which, by differential splicing, is already the basis for extensive functional modulation.

6.4 Downregulation of Nuclear Oncogenes

The oncogene c-*fos* is induced by the activation of adenylyl cyclase through several CRE sites located in different regions of the promoter, although the contribution of the site at position -60 to cAMP-inducibility is major (Sassone-Corsi et al. 1988a; Fisch et al. 1989; Berkowitz et al. 1989). cAMP induction of c-*fos* is followed by a rapid decrease in transcriptional rate, reminiscent of the downregulation observed after serum stimulation (Bravo et al. 1987; Verma and Sassone-Corsi 1987). Fos protein is known to be responsible for the *fos* downregulation after serum stimulation by a mechanism of negative feedback autoregulation exerted on the serum responsive element centered at position -300 of the promoter (Sassone-Corsi et al. 1988b). Fos failed in downregulating the cAMP-induced c-*fos* transcription, making clear that the decrease in transcriptional rate after various stimulatory events could be mediated by different effectors (Foulkes et al. 1991b). It has been shown, in fact, that CREM products are able to downregulate c-*fos* cAMP-induced transcription, whereas are not involved in the downregulation after serum induction. It is possible that this mechanism might have a physiological significance; indeed, the downregulation which follows the induction of c-*fos* expression in the supraoptic nucleus in the brain after osmotic stimulation is paralleled by a sharp increase in CREM antagonists (Mellström et al. 1993). Experiments involving antisense CREM indicated that endogenous CREM could be blocked, thus allowing enhanced cAMP-inducibility and an increased basal c-*fos* transcription level (Foulkes et al. 1991b). Similar experiments conducted on the c-*jun* gene, which is naturally downregulated by cAMP treatment of cultured cells, indicate that CREM is likely to be a molecular effector of this physiological event (E. Benusiglio, personal communication).

7 *Trans*-Activation by Viral Products via the CRE/ATF Binding Proteins

7.1 The ATF Family

Several cDNAs encoding different polypeptides which bind to an ATF site have been cloned and partially characterized (Hai et al. 1989). Screening of a HeLa cell cDNA library revealed the presence of a large family of genes encoding proteins with high homology in the DNA binding domain. Two ATF genes have been characterized in more detail, ATF1 and ATF2. ATF2 is homologous to CRE-BP1, a gene cloned from a brain cDNA library (Maekawa et al. 1989) and to XBP-1 (Kara et al. 1990), cloned from a B-cell specific cDNA library. Despite the high degree of homology among ATF gene products, their combinatorial association is not always possible. Thus, while ATF2 and ATF3 form heterodimers, ATF1 does not dimerize with ATF2 or ATF3. Another noteworthy ATF gene is ATF5, which contains a DNA binding domain with high homology to Fos, suggesting, as in the case of ATF2, possible heterodimerization with nuclear targets of the PKC pathway. All ATF genes characterized to date (Hai et al. 1989) show ubiquitous expression, suggesting that their intracellular levels are unlikely to be major determinants of their activity. Also as with CREB, the ones which have been characterized have trans-regulatory potential and behave as positive activators (Liu and Green 1990).

7.2 Viral Induction via ATF/CRE Sequences

The identification of the CRE as the element necessary for transcriptional activation by elevated cAMP levels was paralleled by the characterization of a similar sequence (the ATF site) to be important in several early promoters of adenovirus (Lin and Green 1988; Sassone-Corsi 1988). ATF sites are responsible for the cAMP-inducibility of some of these promoters. In addition, it was shown that the ATF site in the E4 promoter was crucial for transcriptional activation by the adenovirus E1A protein as well as capable of conferring E1A-inducibility to heterologous promoters (Lee et al. 1989). Interestingly, a number of CRE-containing promoters of cellular genes are inducible by both cAMP and E1A. Recent experiments by several laboratories have shown that E1A-inducibility of CRE/ATF sites is mediated by a specific member of the CRE binding factor family, ATF2 (CREB2, CRE-BP1; Flint and Jones 1991; Lillie and Green 1989; Liu and Green 1990). By fusing specific domains of ATF2 to the DNA binding domain of GAL4 or c-*myb* it was shown that the N-terminal region of ATF2, which contains a putative zinc-finger structure, is both necessary and sufficient for activation by E1A (Flint and Jones 1991; Liu and Green 1990). Interestingly, the N-terminal region of ATF2 can also mediate transcriptional activation by the product of the retinoblastoma gene (Rb; Kim et al. 1992). Substitution of the cysteine residues in the finger structure strongly decreases activation by E1A (Flint and Jones 1991). Studies using GAL4-E1A fusion proteins have also shown that E1A contains a powerful transcription activation domain (Lillie and Green 1989). Therefore, the current model for activation of CRE sites by E1A involves ATF2 as a promoter-bound receptor for the E1A protein, thereby bringing the activation domain of E1A in the proximity of the transcription initiation complex. However, since no direct protein contacts have been demonstrated between ATF2 and E1A, an intermediate factor might act as an adaptor between the two proteins.

Viral induction via CRE sequences does not seem to be limited to the products of adenoviruses. Du and Maniatis (1992) have shown that the element responsible for viral induction of the human interferon-β (HuIFN-β) gene overlaps with a CRE. Mutations in the CRE that diminished binding by ATF2 in vitro also decrease viral induction in vivo. In addition, multiple copies of the virus-inducible element confer both virus and cAMP inducibility to a heterologous promoter. However, the CRE was not sufficient for viral inducibility of HuIFN-β, and additional flanking sequences have been shown to be indispensable.

Another example of viral activation operating through CREs comes from studies on the induction of the HTLV-I LTR of human T-cell leukemia virus (HTLV-I) by the HTLV-I *tax* protein. It was shown that this induction is mediated by three 21-bp repeats present in the LTR (Tan et al. 1989). All three repeats contain a CRE consensus sequence, which are essential for LTR induction by *tax*, and which are able to bind CREB and CREM in vitro (Laoide et al. 1993; Beimling and Moelling 1992). As in the previous case of the IFN-β gene, additional flanking sequences are also required for *tax* induction of the LTR (Fujisawa et al. 1989). Moreover, the *trans*-activation potential of neither CREB, ATF1, nor ATF2 (fused to a GAL4 DNA binding domain) was activated by *tax* in transient cotransfections (Flint and Jones 1991). Thus, the precise mechanism of induction remains unclear.

The most striking example of viral induction through a CRE comes from recent experiments by Maguire et al. (1991). They have shown that induction of the hepatitis B virus (HBV) enhancer element is mediated by a CRE-like sequence. This sequence fails to bind either CREB or ATF2 in vitro. However, when the HBV activator pX was included in the reactions, both CREB and ATF2 efficiently bound to the HBV CRE. Direct protein-protein interactions between pX and CREB or ATF2 have also been demonstrated. Thus, the ability of pX to interact with cellular transcription factors alters their DNA binding specificity, and may ultimately modify the repertoire of genes expressed during viral infection.

8 Physiological Importance of CRE-Binding Proteins

Although the results described above clearly demonstrate that CRE binding factors are important for cAMP-mediated transcriptional regulation in cultured cells, not much has been demonstrated about the specific physiological roles for these proteins. This is of importance because the crucial role played by the variations in cAMP levels in neuroendocrine regulations.

8.1 CREB Function in Pituitary Development

An interesting first clue for a physiological function of CREB came from experiments using transgenic mice that expressed a CREB mutant which cannot be phosphorylated by PKA (Struthers et al. 1991). Since cAMP serves as a mitogenic signal for the somatotroph cells of the anterior pituitary, the mutant cDNA was placed under the control of the somatotroph-specific promoter of the growth hormone gene. The pituitary glands of transgenic mice expressing this construct were atrophied and were deficient in somatotroph cells. Moreover, the transgenic mice exhibited a dwarf phenotype. No other cell type in the pituitary was influenced by expression of the transgene. These effects might arise from repression of genes involved in proliferation and pituitary-specific gene expression, such as c-*fos* and GHF1/Pit-1, although the expression of these genes was not analyzed in the transgenic animals. It is noteworthy that the block of CREB function by the dominant repressor generated a transgenic phenotype equivalent to the one obtained by targeted cell death of the somatomammotrophs (Borrelli et al. 1989). This could be an indication that CRE binding proteins are likely to have pivotal functions in the normal pituitary development.

8.2 CREM Function in Brain

Changes in intracellular levels of cAMP constitute a major regulatory mechanism of signal transduction in the CNS. To date, several nuclear effectors of this pathway have been characterized, although their functional relevance in brain has been unclear because of their widespread distribution and their constant expression. We have reported the specific and anatomically distinct expression of the antagonist isoforms of the CREM gene in adult rat brain

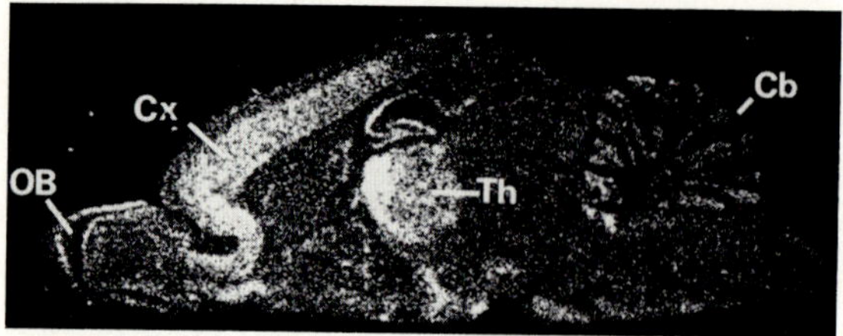

Fig. 4. In situ distribution of CREM transcripts in adult rat brain. Parasagittal sections were hybridized with ^{35}S-labeled antisense riboprobes (Mellström et al. 1993). The probe used for this experiment reveals the distribution of all CREM transcripts. *Cx*, Cerebral cortex; *Cb*, cerebellum; *OB*, olfactory bulb; *Th*, thalmus

and the rapid induction of the α and β isoforms in supraoptic neurons upon physiological stimulation (Mellström et al. 1993). All known CREM isoforms are represented in total brain RNA after PCR amplification (Foulkes et al. 1991a). However, while more quantitative techniques such as RNase protection confirmed the presence of both activators and repressors transcripts, in situ hybridization analysis shows that in neural tissues the antagonist isoforms have a well defined distribution pattern (Fig. 4). In contrast, the activator CREM isoforms, which include the glutamine-rich domains, in common with CREB, have a more diffuse and general distribution. A major point is that CREM differs from the other members of the CRE/ATF family in that specific isoforms are induced upon physiological stimulation. To date, genes of the CRE/ATF class have been described as noninducible (for reviews see Habener 1990; Borrelli et al. 1992). Osmotic stimulation resulted in a differential accumulation of the two antagonist isoforms CREMα and CREMβ, but no change in CREMγ or the activator CREM. Consistent with previous reports (for review see Habener 1990), no induction was observed for the activator CREB.

The induction of several genes in the supraoptic nucleus upon osmotic stimulation has been described previously, including the early response gene c-*fos* which has been shown to undergo a rapid and transient induction (Sherman et al. 1986; Carter and Murphy 1990; Sharp et al. 1991). Since CREM antagonists are able to negatively transregulate the activity of the c-*fos* promoter (Foulkes et al. 1991b), the temporal correlation between the onset of induction of CREMα and -β and the decrease in c-*fos* transcript in supraoptic neurons would suggest a role for CREM antagonists as downregulators of c-*fos* early induction in these neurons. Although our data fall short of unambiguously demonstrating such a role for CREM, they do provide a stimulating basis for future investigations.

A remarkable aspect of the distribution of CREM antagonists in brain is the high level of expression in the anterior thalamic nuclei (Fig. 4). This region, forming a part of the forebrain limbic system, receives input from the hippocampus and the mammillary body of the hypothalamus and projects mainly to the cingulate cortex. This anatomical circuit has been associated with memory and integration of emotions. The significance of CREM expression in the anterior thalamus is unknown, but since this area has been reported to show no induction of early response genes after brain stimulation (Morgan et al. 1987; Sagar et al. 1988; Bullitt 1989), it is tempting to speculate that the high basal expression of CREM antagonists could in part account for this phenomenon. In this respect, it is noteworthy that induction of c-*fos* in the thalamus after peripheral nociceptive or convulsive stimulation occurs in nuclei of the central, midline, and ventral thalamic complexes (Sagar et al. 1988; Bullitt 1989), which in general show a weak hybridization signal for CREM.

A second important observation is the presence of CREM antagonists in almost all the motor nuclei of the brain stem, while sensory nuclei are generally negative. Exceptions to this are the superior olive, which is associated with auditory perceptions, and the mesencephalic trigeminal nucleus, equivalent to the dorsal root ganglia related to propioceptive sensory information, in which CREM transcripts are present. Conversely, CREM expression in motor nuclei includes the somatic motor nuclei – occulomotor, trochlear, abducens, and hypoglossal – as well as the special visceral, trigeminal, and facial nuclei and the general visceral motor nucleus of the vagus. Other positive motor nuclei are the red nucleus, the deep cerebellar nuclei, and the pontine nucleus (Mellström et al. 1993).

Expression of CREM antagonist isoforms in several hypothalamic nuclei associated with homeostatic regulation is also noticeable. Such is the case for magnocellular neurons in the supraoptic hypothalamic nuclei which respond to osmotic stimulation by the differential temporal induction of two of the antagonist isoforms, CREMα and CREMβ. Other functionally related brain areas in which the CREM antagonists are expressed include nuclei involved in visual processing: the suprachiasmatic nucleus, the dorsolateral geniculate nucleus, the lateroposterior thalamic nucleus and the medial terminal nucleus of the accessory optic tract. The latter is supposedly involved in entrainment of endocrine rhythms by light, and fine adjustment of head-eye coordination. Moreover, the presence of CREM in the pineal gland also points to a possible role for CREM in the processing of visual information and the establishment of circadian rhythms.

These findings are a further demonstration of the physiological importance of the CREM gene among the CRE/ATF family of factors. The discovery of the anatomically specific pattern of expression of distinct CREM isoforms, together with their potential for inducibility, sheds new light on the mechanisms whereby cAMP regulates gene expression in the brain.

8.3 Expression of the CRE Binding Protein During Spermatogenesis

Recently a number of reports have suggested a role for CRE binding proteins in spermatogenesis. This was not unexpected since the metabolism of Sertoli's and Leydig's cells, the somatic cell-types that direct the maturation of germinal cells, is regulated by the pituitary gonadotropins follicle-stimulating hormone (FSH) and luteinizing hormone, which in turn activate the adenylyl cyclase pathway. The most striking example of differential regulation of CRE binding proteins during spermatogenesis comes from recent studies on CREM (Foulkes et al. 1992). By studying the expression of the CREM gene during spermatogenesis we observed a novel isoform, CREMτ, which is generated by coordinate insertion of two glutamine rich domains in the repressor isoform CREMβ (see Fig. 2). As a consequence of these insertions, CREM behaves as a transcriptional activator. We found an abrupt switch in CREM expression during spermatogenesis. Premeiotic germ cells express only the repressor forms in low amounts, while from the pachytene spermatocyte stage onwards CREMτ is expressed uniquely and in very high amounts. CREMτ RNA is expressed only in spermatocytes and spermatids, while in Sertoli's and Leydig's cells the preswitch pattern is observed (Foulkes et al. 1992). The CREMτ protein is readily detected in spermatids by using CREM-specific antibodies (Fig. 5; Delmas et al. in preparation). Surgically removing the pituitary in vivo caused the reappearance of the preswitch pattern. By reinjection of different hormones into the testis of these animals it was shown that the developmentally regulated switch of CREM is mediated by FSH (Foulkes et al. 1993). Strikingly, the induction of CREMτ does not occur at the transcriptional level but seems to be mediated by FSH-dependent usage of an alternative polyadenylation site, thereby dramatically enhancing transcript stability (see Fig. 3; Foulkes et al. 1993). These results suggest an important role for CREMτ during spermatogenesis.

In the case of CREB the results are somewhat less striking. It was shown by Waeber et al. (1991) that CREB is highly expressed in the haploid round spermatids of rat testis. However, since the antibody used in these studies was raised against a peptide from a region in CREB that is highly homologous to CREM, it cannot be excluded that the nuclear staining results from cross-reaction to CREM. In addition to nuclear localized full-length CREB or CREM, a truncated form of CREB (CREBW), resulting from alternative splicing was abundant in the cytoplasm of germinal cells. Since this form lacks the bZip domain and the nuclear localization signal, it might perform a

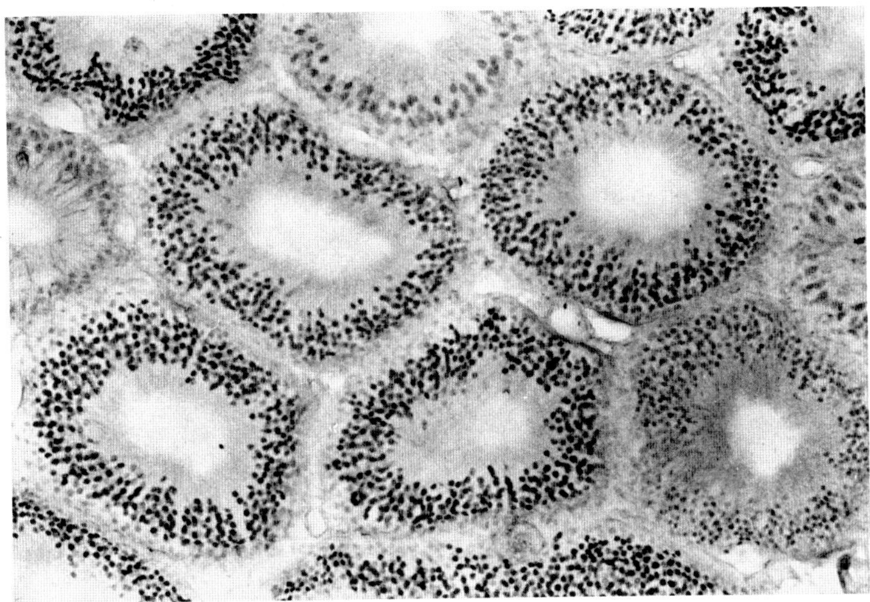

Fig. 5. Peroxidase staining of rat seminiferous tubules showing expression of the CREM protein in the spermatids. The CREM antibody used for this experiment was prepared against a bacterially produced CREMτ protein. Note the differential intensity of the staining in the various tubules, indicating that CREM expression is developmentally regulated

distinct function in the cytoplasm. CREB mRNA was cyclically expressed in Sertoli's cells at stages coinciding with a maximal response of adenylate cyclase to FSH. In addition, Ruppert et al. (1992) demonstrated that the expression CREB and two novel isoforms of CREB, CREBγ and -αγ, was induced in primary spermatocytes after commencement of spermatogenesis. As with CREBW, CREBγ and -αγ encode truncated proteins lacking the bZip domain, the precise role of CREB in spermatogenesis remains to be determined.

9 Cross-Talk at the Nuclear Level

9.1 The PKC-Dependent Pathway

In the PKC-dependent pathway, following the binding of ligands to their cognate receptors, inositol phospholipids are hydrolyzed to generate inositol 1,4,5-triphosphate (IP_3), and DAG, which leads to the activation of PKC (Berridge 1987). In turn, activated PKC phosphorylates several proteins which serve as mediators in the processes of cell proliferation, growth control, and gene regulation.

Transcription factors are among the nuclear targets for PKC transduction pathway (Deutsch et al. 1988; Gilman 1987; Comb et al. 1986; Roesler et al. 1988; Ziff 1990; Binétruy et al. 1991; Pulverer et al. 1991). The nuclear factor AP-1 is phorbol ester-inducible and is constituted by the products encoded by the members of the *jun* gene family, including c-*jun*, *junB*, and *junD* (Bohmann et al. 1987; Angel et al. 1988; Ryder et al. 1988; Lamph et al. 1988; Rauscher et al. 1988; Sassone-Corsi et al. 1998c; Hirai et al. 1989). Jun proteins can homodimerize or heterodimerize with proteins of the Fos family (c–Fos, FosB, Fra-1, and Fra-2; Chiu et al. 1988; Sassone-Corsi et al. 1988c; Zerial et al. 1989; Nishina et al. 1990; Busch and Sassone-Corsi 1990). These dimers bind to TREs present in the promoters of various genes (e.g., metallothionein and collagenase genes), thereby regulating their expression in response to TPA (reviewed in Vogt and Bos 1989; Woodgett 1990).

The *jun* and *fos* genes are members of the immediate early-response class, whose expression is rapidly and transiently induced through intracellular pathways activated by extracellular stimuli (Kruijer et al. 1984; Ryder et al. 1988; Lamph et al. 1988). Functional specificity among the various members of this gene family is likely to be determined by their differential distribution and transcriptional inducibility (Ryder et al. 1988; 1989; Hirai et al. 1989; Naranjo et al. 1991; Mellström et al. 1991). Therefore, the products of these genes act as cell-specific nuclear third messengers, converting cytoplasmic signals into changes in gene expression.

9.2 Cross-Talk Between the PKA and PKC Pathways

Given the many components shared by PKA and PKC pathways, together with the remarkable diversity of agents which can stimulate these pathways, it can be predicted that cross-talk exists between the two signaling systems; indeed, this has been demonstrated in various cases (Cambier et al. 1987; Yoshimasa et al. 1987). Although the difference between TPA and cAMP responsive promoter elements is only one nucleotide (TRE = TGACTCA; CRE = TGACGTCA), they seem to mediate induction only by their respective agonists (Sassone-Corsi et al. 1990). However, this similarity between the nuclear target sequences of the PKA and PKC pathways is striking and suggests possible cross-talk in which CRE binding proteins bind TREs and AP-1 binds CREs.

We have described that the CREM antagonists are capable of negatively modulating the transcriptional activity elicited by Jun/AP-1 (Masquilier and Sassone-Corsi 1992). This phenomenon constitutes a cross-talk between signal transduction pathways at the transcriptional level. Indeed, a canonical TRE can be recognized by two CRE binding proteins, CREB and CREM.

CREM proteins bearing either DNA binding domain are able to bind to TRE sequences. In addition, we observed a dramatic downregulation of the transactivation potential elicited by all Jun proteins (c-Jun, JunB, JunD and v-Jun). There is evidence that the downregulation is likely to be obtained by occupation of the TRE by CREM dimers, since CREM proteins do not heterodimerize with Jun or Fos. Importantly, downregulation is obtained already at a 1:2 ratio of the transfected expression vectors for CREM and Jun, suggesting that CREM binding for the TRE is comparable to Jun. We have found that there are no significant differences between the various CREM proteins in the negative regulation of Jun-mediated transactivation, paralleling the observation that both CREM DNA binding domains confer similar binding activity to the respective CREM proteins (Laoide et al. 1993).

Interestingly, the phosphorylation domain (KID) of CREM is dispensable for the downregulatory function. Infact, CREM truncated proteins which lack the KID region efficiently repress Jun-mediated transactivation. In addition, coexpression of the catalytic subunit of the PKA does not affect the negative regulatory effect of both CREMα and CREMβ. Taking into account these results, and considering that the truncated CREM proteins efficiently bind DNA (Laoide et al. 1993), it seems clear that the negative function of CREM over Jun is almost exclusively due to the occupation of the TRE by the CREM dimer.

One interesting aspect of these results is the fact that transcription factors which could be considered as targets of different signal transduction pathways could affect each others function. It was already shown that Fos-Jun heterodimers bind and activate transcription from CREs (Sassone-Corsi et al. 1990; Hai and Curran 1991). Together with the repression by CREM of the transcriptional activity elicited by Jun/AP1, these observations demonstrate the complexity of the molecular mechanisms of gene regulation and their links with intracellular signal transduction. Thus, there is growing evidence that cross-talk exists at many levels between the two signal transduction pathways.

Interplay may also occur due to promiscuous dimerization among TRE and CRE binding proteins. For example, heterodimerization of ATF2 and Jun switches the specificity of DNA binding with respect to the ATF2 dimer from the CRE sequence to a TRE sequence, thus representing a case of cross-talk between two members of distinct signaling systems (Macgregor et al. 1990; Benbrook and Jones 1990; Ivashkiv et al. 1990; Hai and Curran 1991).

Interestingly, the PKA signal transduction pathway also seems to regulate AP-1 activity through an accessory protein. The characterized inhibitor, IP-1, is itself phosphorylated and thereby inactivated by PKA (Auwerx and Sassone-Corsi 1991). This is reminiscent of the regulation of NFκB by IκB, in which IκB plays an anchoring role (Baeuerle and Baltimore 1988). The protein, IP-1, is present both in the cytoplasm and nucleus of cells and reduces AP-1 complex formation with DNA in a quick and phosphorylation-dependent fashion. The IP-1 protein appears to be very unstable. IP-1 is regulated by phosphorylation and only in its nonphosphorylated form exerts an inhibitory activity on AP-1 DNA binding. Initial purification attempts showed that the protein to have a molecular weight of approximately 43 kDa. IP-1 itself is subject to a complex regulation. Infact, IP-1 activity was shown to be modulated after activation of several signal transduction pathways, including PKC, PKA and Ca^{2+}/calmodulin-dependent kinase pathways, as well as after serum stimulation of cells (Auwerx and Sassone-Corsi 1992). These data obtained after in vivo stimulation of cells with various agents are in good agreement with previous data showing that (a) IP-1 is inactivated by phosphorylation by PKA in vitro and (b) AP-1 function is activated by the PKA, although c-Jun is not phosphorylated by this kinase (de Groot and Sassone-Corsi 1992). Furthermore, the induced differentiation of cultured cells also appeared to influence IP-1 activity, since treatment of P19 EC cells caused an enhanced AP-1 DNA binding and a correlated reduction of IP-1 activity (Auwerx and Sassone-Corsi 1992).

The AP-2 transcription factor recognizes a site distinct from the TRE or CRE binding sites. The AP-2 site appears to be a target for both the cAMP-dependent and PKC signal transduction pathways and may thus constitute a potential site for cross-talk (Imagawa et al. 1987). However, no direct evidence is available indicating the specificity of AP-2 function. In particular, it is unclear whether AP-2 alone can integrate signals from the two pathways or if it requires additional factors.

10 Conclusion

The large number of CRE binding proteins reveals the complexity of the cellular response to cAMP and possibly suggests the requirements of cell-specificity and potential cross-talk mechanisms with the PKC pathway. The CREM gene constitutes a paradigm that represents another level of complexity. Its modularity of function, which is mediated by alternative and cell-specific splicing events, is an example of the versatility that the cell must accomplish in order to permit normal and regulated cell growth in response to several stimuli. The generation of CREB/CREM-deficent animals by homologous recombination will be an important step in the determination of the precise roles played by the different CRE binding proteins in vivo.

References

Acland P, Dixon M, Peters G, Dickson C (1991) Subcellular fate of the Int-2 oncoprotein is determined by choice of initiation codon. Nature 343: 662–665

Andrisani OM, Hayes TE, Roos B, Dixon JE (1987) Identification of the promoter sequences involved in the cell-specific expression of the rat somatostatin gene. Nucleic Acids Res 15: 5715–5728

Angel P, Allegretto EA, Okino ST, Hattori K, Boyle WJ, Hunter T, Karin M (1988) Oncogene jun encodes a sequence-specific trans-activator similar to AP-1. Nature 332:166–171

Auwerx J, Sassone-Corsi P (1991) IP-1: A dominant inhibitor of Fos/Jun whose activity is modulated by phosphorylation. Cell 64: 983–993

Auwerx J, Sassone-Corsi P (1992) AP-1 (Fos-Jun) regulation by IP- 1: effect of signal transduction pathways and cell growth. Oncogene 7: 2271–2280

Baeuerle PA, Baltimore D (1988) IκB: A specific inhibitor of the NFκB transcription factor. Science 242: 540–546

Beimling P, Moelling K (1992) Direct interaction of CREB protein with 21 bp Tax-response elements of HTLV-I LTR. Oncogene 7: 257–262

Benbrook DM, Jones NC (1990) Heterodimer formation between CREB and Jun proteins. Oncogene 5: 295–302

Berkowitz LA., Riabowol KT, Gilman MZ (1989) Multiple sequence elements of a single functional class are required for cyclic AMP responsiveness of the mouse c-*fos* promoter. Mol Cell Biol 9: 4272–4281

Berridge MJ (1987) Inositol trisphosphate and diacylglycerol: two interacting second messengers. Ann Rev Biochem 56:159–193

Binétruy B, Smeal T, Karin M (1991) Ha-Ras augments c-Jun activity and stimulates phosphorylation of its activation domain. Nature 351:122–127

Bohmann D, Bos TJH, Admon A, Nishimura T, Vogt PK, Tjian R (1987) Human proto-oncogene c-jun encodes a DNA binding protein with structural and functional properties of transcription factor AP-1. Science 238:1386–1392

Borrelli E, Heyman R, Arias C, Sawchenko P, Evans RM (1989) Transgenic mice with inducible dwarfism. Nature 339: 538–541

Borrelli E, Montmayeur JP, Foulkes NS, Sassone-Corsi P (1992) Signal transduction and gene control: the cAMP pathway. Critical Rev Oncogenesis 3: 321–338

Bravo R, Neuberg M, Burkhardt J, Almendral J, Wallich R, Muller R, (1987) Involvement of common and cell-type specific pathways in c-fos gene control: stable induction by cAMP in macrophages. Cell 48: 251–260

Bullitt E (1989). Induction of c-*fos*-like protein within the lumbar spinal cord and thalamus of the rat following peripheral stimulation. Brain Res 493: 391–397

Bush SJ, Sassone-Corsi P (1990) Dimers, leucine zippers and DNA binding domains. Trends Genet 6: 36–40

Cambier JC, Newell NK, Justement LB, McGuire JC, Leach KL, Chen ZZ (1987) Ia binding ligand and cAMP stimulates nuclear translocation of PKC in β lymphocytes. Nature 327: 629–632

Carter DA, Murphy D (1990) Regulation of c-*fos* and c-*jun* expression in the rat supraoptic nucleus. Mol Cell Neurobiol 10: 435–446

Chiu R, Boyle WJ, Meek J, Smeal T, Hunter T, Karin M (1988) The c- fos protein interacts with c-Jun/AP-1 to stimulate transcription of AP-1 responsive genes. Cell 54: 541–552

Comb M, Birnberg NC, Seasholtz A, Herbert E, Goodman HM (1986) A cyclic-AMP and phorbol ester-inducible DNA element. Nature 323: 353–356

Courey AJ, Tjian R, (1989) Analysis of Sp1 in vivo reveals multiple transcriptional domains, including a novel glutamine activation motif. Cell 55: 887–898

Dash PK, Karl KA, Colicos MA, Prywes R, Kandel ER (1991) cAMP response element-binding protein is activated by Ca^{2+}/calmodulin-as well as cAMP-dependent protein kinase. Proc Natl Acad Sci USA 88: 5061–5065

deGroot RP, Sassone-Corsi P (1992) Activation of Jun/AP-1 by protein kinase A. Oncogene 7: 2281–2286

deGroot RP, den Hertog J, Vandenheede JR, Goris J, Sassone-Corsi P (1993) Multiple and cooperative phosphorylation events regulate the CREM activator function. EMBO J 12: 3903–3911

Delegeane A, Ferland L, Mellon PL (1987) Tissue specific enhancer of the human glycoprotein hormone α-subunit gene: dependence on cyclic AMP-inducible elements. Mol Cell Biol 7: 3994–4002

Delmas V, Laoide BM, Masquilier D, de Groot RP, Foulkes NS, Sassone-Corsi P (1992) Alternative usage of initiation codons in mRNA encodins the cAMP-responsive-element modulator (CREM) generates regulators with opposite functions. Proc Natl Acad Sci USA 89: 4226–4230

Descombes P, Schibler U (1991) A liver-enriched transcriptional activator protein, LAP, and a transcriptional inhibitory protein, LIP, are translated from the same messenger RNA Cell 67: 569–579

Deutsch PJ, Hoeffler JP, Jameson JL, Habener JF (1988) Cyclic AMP and phorbol ester-stimulated transcription mediated by similar DNA elements that bind distinct proteins. Proc Natl Acad Sci USA 85: 7922–7926

Du W, Maniatis T (1992) An ATF/CREB binding site protein is required for virus induction of the human interferon β gene. Proc Natl Acad Sci USA 89: 2150–2154

Fisch T, Prywes R, Simon MC Roeder RG (1989) Multiple sequence elements in the c-*fos* promoter mediate induction by cAMP. Genes Dev 3: 198–211

Flint KJ, Jones NC (1991) Differential regulation of three members of the ATF/CREB family of DNA-binding proteins. Oncogene 6: 2019–2026

Foulkes NS, Sassone-Corsi P (1992) More is better: activators and repressors from the same gene. Cell 68: 411–414

Foulkes NS, Borrelli E, Sassone-Corsi P (1991a) CREM gene: use of alternative DNA binding domains generates multiple antagonists of cAMP-induced transcription. Cell 64: 739–749

Foulkes NS, Laoide BM, Schlotter F, Sassone-Corsi P (1991b) Transcriptional antagonist CREM down-regulates c-*fos* cAMP-induced expression. Proc Natl Acad Sci USA 88: 5448–5452

Foulkes NS, Mellström B, Benusiglio E, Sassone-Corsi P (1992) Developmental switch of CREM function during spermatogenesis: from antagonist to transcriptional activator. Nature 355: 80–84

Foulkes NS, Schlotter F, Pévet P, Sassone-Corsi P (1993) Pituitary hormone FSH directs the CREM functional switch during spermatogenesis. Nature 362: 264–267

Fujisawa J-I, Toita M, Yoshida M (1989) A unique enhancer element for the trans-activator ($p40^{tax}$) of human T-cell leukemia virus type I that is distinct from cyclic AMP- and 12-O-tetradacanoylphorbol-13-acetate-response elements. J Virol 63: 3234–3239

Gaire M, Chatton B, Kedinger C (1990) Isolation and characterisation of two novel, closely related ATF cDNA clones from Hela cells. Nucleic Acids Res 18: 3467–3473

Gaspar ML, Meo T, Bourgarel P, Guenet JL, Tosi M (1991) A single base deletion in the Tfm androgen receptor gene creates a short-lived messenger RNA that directs internal translation initiation Proc Natl Acad Sci USA 88: 8606–8610

Gilman AG.(1987) G Proteins: Transducers of Receptor-generated Signals. Ann Rev Biochem 86: 615–649

Ginty DD, Glowacka D, Bader DS, Hidaka H, Wagner JA (1991) Induction of immediate early genes by Ca^{2+} influx requires cAMP-dependent protein kinase in PC12 cells. J Biol Chem 266:17454–17458

Gonzalez GA, Montminy MR (1989) Cyclic AMP stimulates somatostatin gene transcription by phosphorylation of CREB at ser 133. Cell 59: 675–680

Gonzalez GA, Yamamoto KK, Fischer WH, Karr K, Menzel P, Briggs III W, Vale WW, Montminy MR (1989) A cluster of phosphorylation sites on the cAMP-regulated nuclear factor CREB predicted by its sequence. Nature 337: 749–752

Gonzalez GA, Menzel P, Leonard J, Fischer WH, Montminy MR (1991) Characterization of motifs which are critical for activity of the cyclic AMP-responsive transcription factor CREB. Mol Cell Biol 11: 1306–1312

Habener J (1990) Cyclic AMP response element binding proteins: a cornucopia of transcription factors. Mol Endocrinol 4: 1087– 1094

Hagiwara M, Alberts A, Brindle P, Meinkoth J, Feramisco J, Deng T, Karin M, Shenolikar S, Montminy M (1992) Transcriptional attenuation following cAMP induction requires PP-1- mediated dephosphorylation of CREB. Cell 70: 105–113

Hai T-Y, Liu F, Coukos WJ, Green MR (1989) Transcription factor ATF cDNA clones: an extensive family of leucine zipper proteins able to selectively form DNA binding heterodimers. Genes Dev 3: 2083–2090

Hai T-Y, Curran T, (1991) Cross-family dimerization of transcription factors Fos/Jun and ATF/CREB alters DNA- binding specificity. Proc Natl Acad Sci 88: 3720–3724

Hirai SI, Ryseck RP, Mechta F, Bravo R, Yaniv M (1989) Characterization of the junD, a new member of the jun proto- oncogene family. EMBO J 8: 1433–1439

Hoeffler JP, Meyer TE, Yun Y, Jameson JL, Habener JF (1988) Cyclic AMP-responsive DNA-binding protein: structure based on a cloned placental cDNA. Science 242: 1430–1433

Hoeffler JP, Lustbader JW, Chen C-Y (1991) Identification of multiple nuclear factors that interact with cyclic adenosine 3' 5'—monophosphate response element-binding protein and activating transcription factor-2 by protein-protein interactions. Mol Endocrinol 5: 256–266

Hurst HC, Totty NF, Jones NC (1991) Identification and functional characterization of the cellular activating transcription factor 43 (ATF-43) protein. Nucl. Acids Res 19: 4601–4609

Imagawa M, Chiu R, Karin M (1987) Transcription factor AP-2 mediates induction by two different signal-transduction pathways: protein kinase C and cAMP. Cell 51: 251–260

Ivashkiv LB, Liou HC, Kara CJ, Lamph WW, Verma IM, Glimcher LH (1990) mXBP/CRE-BP2 and c-jun form a complex which binds to the cAMP, but not to the 12-O-tetradecanoyl phorbol-13- acetate response element. Mol Cell Biol 10: 1609–1621

Kara CJ, Liou HC, Ivashkiv LB, Glimcher LH (1990) A cDNA for a human cyclic AMP response element-binding protein which is distinct from CREB and expressed preferentially in the brain. Mol Cell Biol 10: 1347–1353

Kim S-J, Wagner S, Liu F, O'Reilly MA, Robbins PD, Green MR (1992) Retinoblastoma gene product activates expression of the human TGF-β2 gene through transcription factor ATF-2. Nature 358: 331–334

Kramer IJM, Koornneef I, de Laat SW, van den Eijnden-van Raaij AJM (1991) TGF-β1 induces phosphorylation of the cyclic AMP responsive element binding protein in ML-CCL64 cells. EMBO J 10: 1083–1089

Krebs EG, Beavo JA (1979) Phosphorylation-dephosphorylation of enzymes. Ann Rev Biochem 48: 923–959

Kruijer W, Cooper JA, Hunter T, Verma IM (1984) Platelet-derived growth factor induces rapid but transient expression of the c-*fos* gene and protein. Nature 312: 711–716

Lamph WW, Wamsley P, Sassone-Corsi P, Verma IM (1988) Induction of proto-oncogene Jun/AP-1 by serum and TPA. Nature 334: 626–631

Laoide BM, Foulkes NF, Schlotter F, Sassone-Corsi P (1993) The functional versatility of CREM is determined by its modular structure. EMBO J 12: 1179–1191

Lee KAW, Fink SJ, Goodman RH, Green MR (1989) Distinguishable promoter elements are involved in transcriptional activation by E1A and cyclic AMP. Mol Cell Biol 9: 4390–4397

Lee CQ, Yun Y, Hoeffler JP, Habener JF (1990) Cyclic-AMP- responsive transcriptional activation involves interdependent phosphorylated subdomains. EMBO J 9: 4455–4465

Leff SE, Rosenfeld MG, Evans RM (1986) Complex transcriptional units: diversity in gene expression by alternative RNA processing. Ann Rev Biochem 55: 1091–1117

Leonard J, Serup P, Gonzalez G, Edlund T, Montminy M (1992) The LIM family transcription factor Isl-1 requires cAMP response element binding protein to promote somatostatin expression in pancreatic islet cells. Proc Natl Acad Sci USA 89: 6247- 6251

Lewin B (1991) Oncogenic conversion by regulatory changes in transcription factors. Cell 64: 303–312

Lillie JW, Green MR (1989) Transcription activation by the adenovirus E1a protein. Nature 338: 39–44

Lin Y, Green MR (1988) Interaction of a common cellular transcription factor, ATF, with regulatory units in both E1A and cyclic AMP inducible promoters. Proc Natl Acad Sci USA 85: 3396–3400

Liou H.-C, Boothby MR, Glimcher LH (1988) Distinct cloned class II MHC DNA-binding proteins recognize the X-box transcription element. Science 242: 69–71

Liu F, Green MR (1990) A specific member of the ATF transcription factor family can mediate transcription activation by the adenovirus E1a protein. Cell 61: 1217–1224

Macgregor PF, Abate C, Curran T (1990) Direct cloning of leucine zipper proteins: Jun binds cooperatively to the CRE with CRE- BP1. Oncogene 5: 451–458

Maekawa T, Sakura H, Kanei-Ishii C, Sudo T, Yoshimura T, Fujisawa J, Yoshida M, Ishii S (1989) Leucine zipper structure of the protein CRE-BP1 binding to the cyclic AMP response element in brain. EMBO J 8: 2023–2028

Maguire HF, Hoeffler JP, Siddiqui A (1991) HBV X protein alters the DNA binding specificity of CREB and ATF-2 by protein-protein interactions. Science 252: 842–844

Masquilier D, Sassone-Corsi P (1992) Transcriptional cross-talk: nuclear factors CREM and CREB bind to AP-1 sites and inhibit activation by Jun. J Biol Chem 267: 22460–22466

McCormick A, Brady H, Theill L, Karin M (1990) Regulation of the pituitary-specific homeobox gene GHF1 by cell-autonomous and environmental cues. Nature 345: 829–832

McKnight SG, Clegg CH, Uhler MD, Chrivia JC, Cadd GG, Correll LA, Otten AD (1988) Analysis of the cAMP-dependent protein kinase system using molecular genetic approaches. Rec Progr Horm Res 44: 307–335

Mellon PL, Clegg CH, Correll LA, McKnight SG (1989) Regulation of transcription by cyclic AMP-dependent protein kinase. Proc Natl Acad Sci USA 86: 4887–4891

Mellström B, Achaval M, Montero D, Naranjo JR, Sassone-Corsi P (1991) Differential expression of the jun family members in rat brain. Oncogene 6: 1959–1964

Mellström B, Naranjo JR, Foulkes NS, Lafarga M, Sassone-Corsi P (1993) Transcriptional response to cAMP in brain: specific distribution and induction of CREM antagonists. Neuron 10: 655–665

Montmayeur JP, Borrelli E (1991) Transcription mediated by a cAMP-responsive promoter element is reduced upon activation of dopamine D2 receprors. Proc Natl Acad Sci USA 88: 3135–3139

Morgan JI, Cohen DR, Hempstead JL, Curran T (1987) Mapping patterns of c-*fos* expression in the central nervous system after seizure. Science 237: 192–197

Naranjo JR, Mellström B, Achaval M, Sassone-Corsi P (1991) Molecular pathways of pain: Fos/Jun mediated activation of a non-canonical AP-1 site in the prodynorphin gene. Neuron 6: 607–617

Nichols M, Weih F, Schmid W, DeVack C, Kowenz-Leutz E, Luckow B, Boshart M, Schütz G (1992) Phosphorylation of CREB affects its binding to high and low affinity sites: implications for cAMP induced gene transcription. EMBO J 11: 3337–3346

Nishina H, Sata H, Suziki T, Sato M, Iba H (1990) Isolation and characterization of fra-2, an additional member of the fos gene family. Proc Natl Acad Sci USA 87: 3619–3623

Nishizuka Y (1986) Studies and perspectives of protein kinase C. Science 233: 305–312

Pulverer BJ, Kyriakis JM, Avruch J, Nikolakaki E, Woodgett JR (1991) Phosphorylation of c-Jun mediated by MAP kinases. Nature 353: 670–674

Rauscher FJ, Cohen DR, Curran T, Bos TJ, Vogt PK, Bohman D, Tjian R, Franza Jr BR (1988) Fos-associated protein p39 is the product of the jun proto-oncogene. Science 240: 1010–1016

Rehfuss RP, Walton KM, Loriaux MM, Goodman RH (1991) The cAMP- regulated enhancer-binding protein ATF-1 activates transcription in response to cAMP-dependent protein kinase A. J Biol Chem 266: 18431–18434

Roesler WJ, Vanderbark GR, Hanson RW (1988) Cyclic AMP and the induction of eukaryotic gene expression. J Biol Chem 263: 9063–9066

Ruppert S, Cole TJ, Boshart M, Schmid E, Schütz G (1992) Multiple mRNA isoforms of the transcription activator protein CREB: generation by alternative splicing and specific expression in primary spermatocytes. EMBO J 11: 1503–1512

Ryder K, Lau LF, Nathans D (1988) A gene activated by growth factors is related to the oncogene v-jun. Proc Natl Acad Sci USA 85: 1487–1491

Sagar SM, Sharp FR, Curran T (1988). Expression of c-*fos* protein in brain: metabolic mapping at the cellular level. Science 240: 1328–1331

Saris CJM, Domen J, Berns A (1991) The pim-1 oncogene encodes two related protein-serine/threonine kinases by alternative initiation at AUG and CUG. EMBO J 10: 655–664

Sassone-Corsi P (1988) Cyclic AMP Induction of early adenovirus promoters involves sequences required for E1A-transactivation. Proc Natl Acad Sci USA 85: 7192–7196

Sassone-Corsi P, Visvader J, Ferland L, Mellon PL, Verma IM (1988a) Induction of proto-oncogene *fos* transcription through the adenylate cyclase pathway: characterization of a cAMP-responsive element. Genes Dev 2: 1529–1538

Sassone-Corsi P, Sisson JC, Verma IM (1988b) Transcriptional autoregulation of the proto-onogene fos. Nature 334: 314–319

Sassone-Corsi P, Lamph WW, Kamps M, Verma IM (1988c) fos- associated cellular p39 is related to nuclear transcription factor AP-1. Cell 54: 553–560

Sassone-Corsi P, Ransone LJ, Verma IM (1990) Cross-talk in signal transduction: TPA-inducible factor Jun/AP-1 activates cAMP responsive enhancer elements. Oncogene 5: 427–431

Sharp FR, Sagar SM, Hicks K, Lowenstein D Hisanaga K (1991) c-*fos* mRNA, Fos and Fos-related antigen induction by hypertonic saline and stress. J Neurosci 11: 2321–2331

Shaw G, Kamen R (1986) A conserved AU sequence from the 3' untranslated region of GM-CSF mRNA mediates selective mRNA degradation. Cell 46: 659–667

Sheng M, McFadden G, Greenberg ME (1990) Membrane depolarization and calcium induce c-*fos* transcription via phosphorylation of transcription factor CREB. Neuron 4: 571- 582

Sheng M, Thompson MA, Greenberg ME (1991) CREB: a Ca^{2+}- regulated transcription factor phosphorylated by calmodulin- dependent kinases. Science 252: 1427–1430

Sherman TG, McKelvy JF, Watson SJ (1986) Vasopressin mRNA regulation in individual hypothalamic nuclei: a northern and in situ hybridization analysis. J Neurosci 6: 1685–1694

Struthers RS, Vale WW, Arias C, Sawchenko PE, Montminy MR (1991) Somatotroph hypoplasia and dwarfism in transgenic mice expressing a non-phosphorylatable CREB mutant. Nature 350: 622–624

Tan T-H, Horikoshi M, Roeder RG (1989) Purification and characterization of multiple nuclear factors that bind to the tax-inducible enhancer within the human T-cell leukemia virus type I long terminal repeat. Mol Cell Biol 9: 1733–1745

Vinson CR, Sigler P, McKnight SL (1989) Scissor-grip model for DNA recognition by a family of leucine zipper proteins. Science 246: 911–922

Verma IM, Sassone-Corsi P (1987) Proto-oncogene *fos*: complex but versatile regulation. Cell 51: 513–514

Vogt PK, Bos TJ (1989) The oncogene Jun and nuclear signalling. Trends Biochem Sci 14: 172–175

Waeber G, Meyer TE, LeSieur M, Hermann HL, Gérard N, Habener JF (1991) Developmental stage-specific expression of cyclic adenosine 3',5'-monophosphate response element-binding protein CREB during spermatogenesis involves alternative exon splicing. Mol Endocrinol 5: 1418–1430

Williams T, Admon A, Luscher B, Tjian R (1988) Cloning and expression of AP-2, a cell-type-specific transcription factor that activates inducible enhancer elements. Genes Dev 2: 1557–1569

Woodgett JR (1990) Fos and Jun: two into one will go. Sem Cancer Biol 1: 389–397

Yamamoto KK, Gonzales GA, Briggs III WH, Montminy MR (1988) Phosphorylation-induced binding and transcriptional efficiency of nuclear factor CREB. Nature 334: 494–498

Yoshimasa T, Sibley DR, Bouvier M, Lefkowitz RJ, Caron MG (1987) Cross-talk between cellular signalling pathways suggested by phorbol ester adenylate cyclase phopshorylation. Nature 327: 67–70

Yoshimura T, Fujisawa J-I, Yoshida M (1990) Multiple cDNA clones encoding nuclear proteins that bind to the Tax-dependent enhancer of HTLV-I: all contain a leucine zipper structure and basic amino acid domain. EMBO J 9: 2537–2542

Zanger UM, Lund J, Simpson ER, Waterman MR (1991) Activation of transcription in cell-free extracts by a novel cAMP- responsive sequence from the bovine CYP17 gene. J Biol Chem 266: 11417–11420

Zerial M, Toschi L, Ryseck RP, Schuermann M, Muller R, Bravo R (1989) The product of a novel growth factor activated gene, FosB, interacts with Jun proteins enhancing their DNA binding activity. EMBO J 8: 805–813

Ziff EB (1990) Transcription factors: a new family gathers at the cAMP response site. Trends Genet 6: 69–72

Editor-in-charge: Prof. H. Grunicke and Prof. H. Schweiger

Rev. Physiol. Biochem. Pharmacol., Vol. 124
© Springer-Verlag 1994

Suppression of *ras* Oncogene-Mediated Transformation

R. SCHÄFER

Contents

Division of Cancer Research, Department of Pathology, University of Zurich, Schmelzberg-
strasse 12, CH-8091 Zurich, Switzerland

1 The Function of *ras* in Normal and Transformed Cells

Ras proteins play a fundamental role in the transduction of stimuli from the cellular environment via cell membrane receptors to the nucleus. Both normal and mutated forms of the *ras* gene products have been shown to be involved in the control of proliferation and differentiation. The *ras* genes were originally identified as the transforming genetic elements in the genomes of the Harvey and Kirsten strains of rat sarcoma viruses. The genome of a mammalian cell harbors the Ha-*ras*-1 and Ki-*ras*-2 proto-on-cogenes as well as several pseudogenes. A third member of the *ras* gene family, designated N-*ras*, was isolated, for which a retroviral derivative is not yet known (for review see Barbacid 1987; Grand and Owen 1991). Ras proteins belong to the superfamily of small molecular weight GTPases comprising about 50 different members (for review see Bourne et al. 1991). The size of *ras* genes varies from the 4.5 kb of c-Ha-*ras*-1 located on human chromosome 11p15-p15.5 to the 50 kb of c-Ki-*ras*-2 mapped in chromosome 12p12.1-pter. The human N-*ras* gene has been assigned to the short arm of chromosome 1 in band p22-p32. In spite of varying intron structures, the coding sequences of the *ras* genes equally consist of four exons encoding a protein product of 21 kDa, known as p21^{ras}. The c-Ki-*ras*-2 gene has two alternative fourth coding exons allowing the synthesis of two isomorphic p21 proteins with a different carboxy-terminal domain. The 85 amino-terminal amino acid residues are identical among mammalian Ras proteins. The following internal 80 amino acid residues share 85% similarity among different *ras* genes, whereas the remaining C-terminal amino acid residues are highly variable except for the terminal four amino acid residues (for review see Barbacid 1987; Grand and Owen 1991).

Cellular oncogenes were identified by transfection of DNA of chemically transformed cells into phenotypically normal recipient cells (Shih et al. 1979; Cooper et al. 1980). Introduction of these genes into preneoplastic murine NIH/3T3 fibroblasts resulted in morphological transformation (focus formation), anchorage-independent proliferation, and tumorigenicity. Later, tumor DNA mediated-transformation of those phenotypically normal indicator cells became a widely used strategy to detect cellular oncogenes. The transforming activity of cellular Ha-*ras*, Ki-*ras*, and N-*ras* genes found in tumors was due to point mutations affecting amino acid codons 12, 13, 59, and 61 (for review see Barbacid 1987; Grand and Owen 1991). Morphologically transformed foci were also induced in NIH/3T3 cells following overexpression of the nonmutated Ha-*ras* gene (Chang et al. 1982). The number of human and experimental tumors screened for *ras* mutations increased significantly, when the laborious NIH/3T3 transfection assays were replaced by direct mutational analysis of the genes using oligonucleotide mismatch hybridization (Bos et al. 1984) or RNase mismatch cleavage (Winter et al. 1985). The incidence of *ras* gene mutations varied greatly in human tumors. High incidences of mutations were found in carcinomas of the pancreas (90%), the colon (50%), and the lung (30%), in thyroid tumors (50%), and in myeloid leukemia (30%; for review see Bos 1989).

Transforming *ras* genes were not only detected by gene transfer experiments in cultured cells. The critical role of mutated (activated) *ras* genes in carcinogenesis was confirmed in experimental cancer systems. Activated c-Ha-*ras* genes were reproducibly detected in skin papillomas and carcinomas of mice treated with dimethylbenzanthracene (Balmain et al. 1984). Mouse mammary carcinomas induced by nitrosomethylurea consistently harbored Ha-*ras* mutations (Sukumar et al. 1983; Zarbl et al. 1985). Exposure to the same carcinogen or gamma-radiation of mice resulted in the formation of thmyic lymphomas carrrying N-*ras* mutations (Guerrero et al. 1984a,b). A comprehensive list of *ras* mutations associated with carcinogen exposure has been reported by Barbacid (1987). More recently, transgene technology has been used to target the expression of activated *ras* genes to defined tissues. N-*ras* or Ha-*ras* oncogenes were cloned under the control of a chimeric immunoglobulin (IgH) enhancer/SV40 promoter, of the mouse mammary tumor virus long terminal repeat (MMTV-LTR), whey acidic protein (WAP) promoter, elastase promoter, and growth hormone releasing factor (GHRF) promoter. Transgenic mice carrying these gene constructs gave rise to neoplasia. The tumors included T-cell lymphomas (N-*ras* gene controlled by IgH-SV40 enhancer/promoter), lung adenocarcinomas (IgH-SV40/Ha-*ras*), mammary carcinomas (MMTV-LTR/Ha-*ras*), and neonatal acinar cell hyperplasia (elastase promoter/Ha-*ras*). Rare mammary tumors were detected in mice carrying WAP/Ha-*ras* transgenes, and no phenotypic changes could be observed in mice with the GHRF/Ha-*ras* transgene (for review see Hanahan 1988; Pattengale et al. 1989; Cardiff et al. 1991).

Ras proteins bind guanine nucleotides (GDP and GTP). The bound GTP is hydrolyzed to GDP and inorganic phosphate. These biochemical properties of p21^{ras} are essential for its biological function. When anti-*ras* antibodies were microinjected into *ras*-transformed NIH/3T3 cells, guanine nucleotide binding was inhibited, and neoplastic transformation was abolished. Mutated Ha-*ras* genes that do no longer bind guanine nucleotides were unable to neoplastically transform NIH/3T3 cells. Transforming alleles of *ras* genes carrying point mutations exhibited a tenfold decrease of GTPase activity together with accumulation of the GTP-bound form of the protein. It was concluded that Ras receives a signal from a molecule located upstream in the signaling pathway and that the activated protein transmits this stimulus to a downstream effector most probably via interaction by its effector region (for review see Barbacid 1987; Grand and Owen 1991; Satoh et al. 1992). The levels of GDP and GTP bound to Ras were determined in intact normal cells. There are various extracellular stimuli able to activate wild-type Ras by enhancing the proportion of GTP-bound protein in different types of cells. These include not only the neoplastic transformation of fibroblast cells by receptor-type or non receptor-type oncogenes (Gibbs et al. 1990; Satoh et al. 1990), but also the stimulation of normal cells by growth factors such as PDGF, EGF, and insulin (Mulcahy et al. 1985; Cai et al. 1990; Medema et al. 1991). Stimulation of T cell, B cell, and mast cell lines by different interleukins and colony-stimulating factors lead to accumulation of Ras-GTP (Satoh et al. 1991; Duronio et al. 1992; Graves et al. 1992; Heidaran et al. 1992). In erythroleukemic cell lines, Ras was activated by erythropoeitin (Torti et al. 1992). Activation of T lymphocytes appeared to be mediated via Ras-GTP (Downward et al. 1990; Graves et al. 1991). However, Ras-GTP is not only involved in signaling events that trigger cellular proliferation. Ras proteins may function as transducers of neuronal differentiation signals in rat PC12 pheochromocytoma cells (Noda et al. 1985; Bar-Sagi and Feramisco 1985), of adipocytic conversion in murine 3T3-L1 fibroblasts (Benito et al. 1991), of steroidogenic differentiation of ovarian surface epithelial cells (Pan et al. 1992), and of mesoderm induction in *Xenopus* embryogenesis (Whitman and Melton 1992). In addition, growth inhibition of epithelial cells by transforming growth factor β is associated with GTP binding to Ras and thus dependent on Ras signal transduction (Mulder and Morris 1992). Expression of v-Ha-*ras* lead to growth arrest in Schwann cells that express a temperature-sensitive mutant of SV40 large T antigen at the restrictive temperature (Ridley et al. 1988). Ras proteins are components of a cascade of signaling events rather than of an unidirectional signal transduction pathway. A number of molecules have been identified which modulate Ras activity by interfering with the state of bound guanine nucleotides (for review see Satoh et al. 1992). The signaling events between a receptor tyrosine kinase located at the cell surface and Ras have very recently been elucidated. Several groups have described physical interactions between the internal domain of ligand-activated kinase receptors and GrB2 protein, which binds to mSos1, a Ras nucleotide-exchange protein. Receptor-activated Sos facilitates GDP to GTP exchange on Ras (for review see McCormick 1993). Active GTP-Ras feeds the signal into a cascade of serine-threonine kinases which send it further to the nucleus (for review see Roberts 1992).

2 Limited Contribution of Mutant *ras* Genes to Malignant Transformation

The expression of transfected *ras* oncogenes in preneoplastic mouse NIH/3T3 cells resulted in morphological transformation, anchorage-independent growth, and tumorigenicity in nude mice. The intrinsic GTPase activity of Ras is greatly diminished upon mutation. Therefore, the transforming activity of the mutated protein is explained by the preferential maintenance of the active, GTP-bound state. Increasing the proportion of Ras-GTP by overexpression of the normal *ras* allele may result in the same phenotype (for review see Barbacid 1987). Since NIH/3T3 cells are immortal and exhibit an aneuploid phenotype, it is conceivable that genetic alterations facilitating the transforming activity of oncogenes have already occurred. The genome of normal diploid embryo fibroblasts provides a more efficient barrier against the deleterious effects of *ras* oncogenes. Neoplastic transformation of rat embryo fibroblasts was not achieved by expression of an activated *ras* gene alone, but required the simultaneous expression of a cooperating oncogene such as *myc* or adenovirus E1A (Land et al. 1983; Ruley 1983). Oncogene cooperation was also demonstrated in vivo (Paterson et al. 1989). Since transformed cell lines obtained after transfection of oncogenes or tumors emerging in oncotransgene-carrying mice were usually not analyzed by karyotyping, important gross genetic alterations might have escaped attention. Tumor formation by transformed Syrian hamster embryo fibroblast required the expression of cooperating *ras* and *myc* oncogenes and the loss of a putative tumor suppressor gene as indicated by consistent chromosomal deletions (Oshimura et al. 1985) or by the loss of regulatory signals from neighboring cells mediated by gap-junctional communication (Land et al. 1986). Growth factor-independent proliferation of murine hematopoietic cells was associated with a great variability in their karyotypes (Vogt et al. 1986). Neoplastic transformation of rat embryo fibroblasts mediated by overexpression of mutated *ras* alone as described in one report (Spandidos and Wilkie 1984) can possibly be explained by similar undetected genetic alterations. Most importantly, although all cells in a given tissue to which expression of oncotransgenes has been targeted provide the necessary transcriptional machinery for their activity, it is only a rare fraction of cells that give rise to malignant growth. Consequently, tumors obtained in mice carrying oncogenes as transgenes were mono- or oligoclonal, suggesting that additional changes had occurred (for review see Pattengale et al. 1989; Cardiff et al. 1991). Similarly, tumors derived after infection of midgestation mouse embryos with recombinant retroviruses carrying *ras* and *myc* oncogenes were of monoclonal origin. Thus, additional alterations were necessary for the realization of the fully malignant phenotype (Compere et al. 1989).

Molecular genetic studies in human tumors support the notion that the malignant state is characterized by a set of genetic alterations including oncogene activation and loss of suppressor gene function. For example, Ki-*ras* genes are not only frequently mutated (for review see Bos 1989), but also play an important functional role in colorectal tumors. Disruption of this oncogene in two human colon carcinoma cell lines by homologous recombination resulted in morphological alterations, loss of anchorage-independent proliferation, and reduction in tumor formation in nude mice (Shirasawa et al. 1993). The mutational activation of a *ras* oncogene did not induce the transformed phenotype in a dominant manner, but rather predisposed a nontumorigenic immortalized cell line to transformation by independent events. This was shown by the replacement in Rat-1 fibroblasts of one of the normal Ha-*ras* alleles by a mutated copy via homologous recombination (Finney and Bishop 1993). The transition from normal colonic epithelium to a malignant tumor capable of shedding metastatic cells appears to involve the functional inactivation of several transformation suppressor genes (Fearon and Vogelstein 1990). Tumor suppressor genes of relevance for both hereditary and sporadic colorectal tumors include the *AFP* gene (Kinzler et al. 1991), *p53* (Baker et al. 1989), and the *DCC* gene (Fearon et al. 1990). The number of genetic lesions, involved in colorectal carinogenesis is still increasing (Peltomäki et al. 1993; Aaltonen et al. 1993). Other human tumors may have sustained a number of mutations in some of those genes as well as in genes located on different chromosomes. The *p53* gene acts as a suppressor of proliferation and neoplastic transformation in human tumor cell lines (Baker et al. 1990; Goyette et al. 1992). Circumstantial evidence suggests that the *DCC* gene which encodes a cellular adhesion protein can also restrain malignant proliferation (Narayanan et al. 1992).

The deletions and/or mutations in tumor suppressor genes found in transformed cells with activated *ras* oncogenes suggest that the constitutively activated signal transduction pathway does not by itself mediate uncontrolled proliferation. The changes occurring in *ras*-expressing cells comprise irreversible mutational events. Recently it was demonstrated that *ras* oncogene expression increased both the frequency of chromosomal aberrations and of spontaneous mutations in the hypoxanthine phosphoribosyl-transferase gene (Baron et al. 1992). The *ras*-mediated transformation is associated with the modulation of gene expression. These alterations include both enhanced expression of cellular genes as well as downregulation of gene activity (for review see Klemenz and Aoyama 1993).

3 Suppression of *ras*-Mediated Neoplastic Transformation

3.1 Cell Fusion Studies

Several lines of evidence have suggested that the neoplastic phenotype of cells expressing *ras* oncogenes can be reversed. Somatic cell hybridization of transformed and normal cells had suggested that normal cells contain transformation-suppressing activity. The suppression of tumorigenicity and of anchorage-independent proliferation appeared to have a genetic basis. Whereas the suppressed hybrids often contained the full chromosomal complement of both parental genomes, reemergence of transformed phenotypes in cell segregants was associated with loss of chromosomes derived from the normal parental genome (for review see Stanbridge 1991). Several groups have reported on the isolation of suppressed somatic cell hybrids between *ras* oncogene-expressing cell lines and normal cells. A nontransformed Chinese hamster embryo fibroblast cell line (CHEF) was fused with a hamster transfectant line carrying the human mutant c-Ha-*ras* gene (Craig and Sager 1985). Selected hybrids exhibited a fibroblastic morphology, had a reduced capacity for anchorage-independent growth and decreased tumor-forming ability in nude mice. Suppressed hybrid clones contained the mutant *ras* gene and expressed its p21 protein product at levels comparable to the transformed parental cells. Somatic cell hybrids of Rat-1 cells transformed with the activated c-Ha-*ras* gene from human EJ bladder carcinoma cells and diploid embryonic rat fibroblasts expressed one-tenth to one-third of the mutant p21^{ras} product as compared to parental cells. The hybrid cells showed the suppressed phenotype. It was concluded that the transforming activity of the activated Ha-*ras* gene can be suppressed at the posttranslational level by the presence of the genome of normal fibroblasts (Griegel et al. 1986). The v-Ha-*ras* oncogene was introduced into a mouse mastocytoma cell line which acquired the capacity to produce interleukin-3 (IL-3) and to initiate tumor growth. When this cell line was fused with its IL-3-dependent, nontransformed parental cell line, hybrid cell lines dependent on IL-3 for proliferation in vitro were isolated. The endogenous IL-3 production was down-regulated in these hybrids. Although the p21^{ras} protein was expressed at high levels, the tumor-forming potential of hybrid cells was reduced, as indicated a by fivefold prolonged latency period (Diamantis et al. 1989). The authors concluded that a tumor suppressor present in mouse mastocytes acts as a negative regulator of IL-3 expression and is responsible for the loss of autonomous growth.

The suppression of *ras*-transformation is not restricted to rodent cell hybrids. When human EJ bladder carcinoma cells, harboring a mutant Ha-*ras* allele, were fused with a normal human fibroblast line, hybrids were isolated that behaved as transformed cells in culture but did not form tumors in nude mice (Geiser et al. 1986). The hybrid cells maintained the suppressed phenotype, even if the level of Ha-*ras* expression was elevated by transfection of additional copies of the oncogene. Results of cell fusion experiments involving different human cell lines expressing either Ha-*ras*, Ki-*ras*, or N-*ras* oncogenes suggested that suppression of tumorigenicity can occur even in the presence of two transforming genes of the *ras* family (Geiser et al. 1989). Human uroepithelial cells immortalized by SV 40 were transfected with the *ras* oncogene derived from EJ bladder carcinoma cells. Tumorigenic transformation of these epithelial cells occurred as a rather rare event. Cell fusion of a tumorigenic transformant, and a nontumorigenic transfectant clone indicated the suppression of tumorigenicity in spite of mutant p21^{ras} expression (Pratt et al. 1992). The ability to block neoplastic transformation mediated by N-*ras* of PA-1 human teratocarcinoma cells was dependent on the duration of cell culture (Krizman et al. 1990). While early-passage PA-1 cells were resistant toward the oncogene, late-passage cells could be transformed with high efficiency. Fusion of low- and high-passage cells showed that the resistance toward transformation is a dominant trait. A *ras*-resistant derivative of PA-1 teratocarcinoma cells cannot be stimulated to grow without anchorage by growth factors added to the medium (Chiao et al. 1991).

The results from cell fusion studies using *ras*-expressing cells suggest that the transformed properties in vitro and in vivo are controlled by unknown tumor suppressor genes. The ability to metastasize may be suppressed as well since the metastatic potential of v-Ha-*ras* expressing rat mammary cells was abolished upon fusion with the nonmetastatic dimethylbenz[a]anthracene-induced parental line. The hybrid clones maintained their ability to form primary tumors (Ichikawa et al. 1992). All hybrid clones continued to express the *ras* oncogene.

To identify chromosomes which carry putative tumor suppressor genes, Yamada et al. (1990) introduced microcells harboring either human chromosomes 1, 11 or 12 tagged with pSV2neo DNA into Kirsten sarcoma-virus transformed NIH/3T3 cells. The growth rate, colony-forming ability in semisolid agar medium and tumorigenicity in nude mice was reduced in Ki-*ras* transformed cells carrying chromosome 1, but not in those cells into which chromosome 11 or 12 was inserted. Recently, growth suppression was shown in human RD rhabdomyosarcoma cells which harbor an activated N-*ras* gene (Hall et al. 1983). A modification of the microcell fusion technique was used that permits transfer of subchromosomal fragments rather than individual chromosomes (Koi et al. 1993). Microcells were prepared from mouse A9 cells harboring human t(11;X) chromosomes as the only exogenous chromosomes tagged with multiple copies of the *neo* gene. The microcells were gamma-irradiated to generate subchromosomal transferable fragments (STF). Microcell-mediated fusion of irradiated chromosomal fragments into A9 cells allowed the propagation and subsequent transfer of small defined parts of this human chromosome. With the help of chromosome-specific DNA markers the precise origin of human chromosomal fragments transferred into A9 cells was determined. When ten different STF specific for chromosome band 11p15 were introduced into RD rhabdomyosarcoma cells, eight microcell fusions gave rise to colonies consisting of flat, enlarged, and elongated cells. These colonies ceased to proliferate 2–4 weeks after microcell fusion. By comparison with STF that did not transmit antiproliferative activity, it was concluded that the suppressive activity is associated with band 11p15.5.

3.2 *ras*-Resistant Cell Lines

Established rodent cell lines have been described that cannot be stably transformed by overexpressed *ras* oncogenes. Introduction of activated *ras* failed to transform REF52 cells. Tumorigenic transformation required collaborating adenovirus E1A or simian virus large tumor (T) antigen oncogenes (Franza et al. 1986; Hirakawa and Ruley 1988). Thus, immortalized REF52 cells resemble normal rat embryo fibroblasts. The resistance toward *ras* transformation indicates that immortalization per se is not a sufficient prerequisite for tumorigenic conversion by this oncogene. In the presence of E1A expression, higher levels of $p21^{ras}$ were accumulated in REF52 cells (Franza et al. 1986). Elevating the level of $p21^{ras}$ in the absence of E1A resulted in cell growth arrest and morphological crisis. A derivative of mouse NIH/3T3 cells, designated EK-3, exhibits a similar resistance toward transformation by a *ras* oncogene alone (Katz and Carter 1986; Katz and Samid 1989). Elevated levels of Ha-*ras* mRNA were insufficient to induce morphological transformation and anchorage-independent proliferation in this cell line. In addition, a number of gene transfer experiments using *ras* oncogenes alone or in combination with a collaborating oncogene have been reported that failed to transform normal human recipient cells. The efficient expression of oncogene products was demonstrated in these cells (Stevenson and Volsky 1986; Sager et al. 1986). The finite life-span of human recipient cells appeared to limit the expression of malignancy, since *ras* oncogenes were able to transform immortalized, nontumorigenic human cells (Hurlin 1989; Seremetis et al. 1989; Wilson et al. 1990; Fry et al. 1990).

4 Phenotypic Revertants Derived from *ras*-Transformed Cells

4.1 Oncogene, Suppressor, and Effector Mutants

Morphologically altered (flat) revertants were obtained from cell lines transformed by the viral *ras* oncogene. Flat revertants were isolated from populations of cells transformed by Kirsten murine sarcoma virus by cell "suicide" techniques. The principle of these methods is that cells were cultured under conditions which favor the proliferation of transformed cells but result in growth arrest of revertant cells. Those cells were protected because of their mitotic inactivity. Transformed cells were mutagenized prior to the selection procedure to increase the frequency of phenotypic reversion. The selective culture conditions included treatment with 5-fluorodeoxyuridine at high density, in methylcellulose (Ozanne and Vogel 1974) or in serum-poor medium (Vogel and Pollack 1974), with concanavalin A (Ozanne and Vogel 1974), with iododeoxyuridine in methylcellulose (Greenberger and Aaronson 1974), and with high temperature (Cho et al. 1976). Revertant cell lines were also isolated from clonal lines of virus-transformed cells at very low frequency by selection with bromodeoxyuridine (Stephenson et al. 1973; Morris et al. 1980; Norton et al. 1984). The loss of transformed characteristics resulted from mutations affecting the expression of the functional viral oncogene (Ozanne and Vogel 1974; Greenberger and Aaronson 1974; Cho et al. 1976; Vogel and Pollack 1974). Other revertant cell lines had sustained mutations in cellular effector genes necessary for the expression of the transformed phenotype (Stephensen et al. 1973; Morris et al. 1980; Norton et al. 1984). The early work on phenotypic revertants has been reviewed comprehensively by Bassin and Noda (1987). The expression of neoplastic transformation may also be modulated by changes in the dosage of the oncogene. Revertants derived from human HT1080 fibrosarcoma cells after mutagenesis and cell "suicide" selection showed an increase in chromosomal ploidy, while the one mutated allele of activated N-*ras* persisted. The level of the mutant $p21^{N\text{-}ras}$ product was therefore decreased in the revertants relative to the parental cell line. Introduction of another copy of the transforming allele resulted in retransformation of the revertants (Paterson et al. 1987).

Cell lines that continue to express the oncogene despite the loss of transformed properties were of particular interest for the study of the mechanisms of oncogene resistance. DT, a clone of NIH/3T3 cells containing two copies of the Ki-MuSV genome, was used for the isolation of flat revertants (Noda et al. 1983). The number of flat colonies resulting from mutations in the v-*ras* gene was minimized by the presence of two integrated copies of the virus in this cell line. When *ras*-transformed DT cells were treated with the mutagen 5'-azacytidine and selected in medium depleted of K^+ and supplemented with ouabain, the *ras*-transformed cells showed a significantly lower efficiency of cloning in the presence of the drug. In contrast, normal NIH/3T3 and other cell lines not expressing activated *ras* genes are relatively insensitive toward ouabain-treatment (Benade et al. 1986; Talbot et al. 1988; Wang et al. 1988). Noda et al. (1983) chose two ouabain-resistant flat colonies for further analysis. These two revertant cell lines had lost several properties associated with *ras* transformation. These properties included a short doubling time, the ability to form colonies in semisolid agar medium and to initiate tumor growth in nude mice. In spite of the expression of the nontransformed phenotype, the cells continued to express p21^{ras} at an unaltered level. Transforming virus particles could be rescued from revertant cell lines. The authors concluded that the two revertant cell lines possess alterations in their cellular genomes that allow them to block the Ki-MuSV-induced transformation. The block occurred at some point distal to the initial interaction of p21 with its target molecules. The revertant cell lines were fused with cell lines transformed by Ha-MuSV, BALB-MuSV, v-*mos*, v-*fes*, v-*fms*, v-*sis*, v-*src* oncogenes, or polyomavirus. The resulting somatic cell hybrids were analyzed for their ability to proliferate in semisolid agar medium. Neoplastic transformation triggered by *ras*, *fes*, and *src* oncogenes was suppressed in cellular hybrids, whereas *mos*-, *fms*-, and *sis*- and polyomavirus-transformed cells were unaffected (Table 1).

Kuzumaki et al. (1989) isolated a flat revertant clone, designated R1, from NIH/3T3 cells transformed by an activated human Ha-*ras* gene after mutagenesis. R1 cells were resistant to the transforming activity of Kirsten sarcoma virus, the cellular mutant Ha-*ras* gene, v-*src*, v-*mos* oncogenes, SV40 large T antigen, or polyoma middle T antigen. To find out whether R1 cells had sustained mutations in cellular genes that interfered with the expression of the neoplastic phenotype, the authors performed cell fusion of R1 and NIH/3T3 cells. Somatic cell hybrids expressed the nontransformed phenotype suggesting the activation of transformation suppressor genes in R1.

Two new revertant clones have been isolated from the *ras*-transformed cell line DT using *cis*-4-hydroxy-L-proline (CHP) as a selective agent (Yanagihara et al. 1990). CHP is able to interfere with the synthesis of proteins that are rich in proline and hydroxyproline such as collagen. Since normal collagen production is necessary for cell growth, CHP acts as a growth inhibitor. Rodent cells transformed by different oncogenes, RNA and DNA tumor viruses or by the carcinogen nitrosomethylurea are more susceptible to the antiproliferative properties of CHP than are their untransformed precursor cells (Ciardiello et al. 1988). Although the two CHP-resistant revertants continued to express the v-Ki-*ras* oncogene and contain rescuable virus, they were anchorage-dependent and only weakly tumo-

Table 1. Phenotypic revertants: pattern of resistance toward oncogene-mediated transformation

Parental cell line	Oncogene	Mode of revertant selection	ras	src	fes	fgr	fms	abl	mos	raf	erb-B2	sis	fos	Pol mT
NIH/3T3	v-K-*ras*	AC, ouabain	+	+	+		−		−			−		−
NIH/3T3	v-K-*ras*	EMS, CHP	+	−	−	−	−		+	+		−	+	
NIH/3T3	v-K-*ras*	EMS, ouabain	+	+	+	+	−		−	−		−	−	
NIH/3T3	c-H-*ras* mutant	EMS, AzaG	+	+					+					+[a]
NIH/3T3	c-H-*ras* mutant	BrdU + irradiation	+					(+)			−	−		
NIH/3T3	LTR-c-H-*ras*	Interferon α/β	+		+			+						[b]
NIH/3T3	c-H-*ras* mutant	Interferon γ	+	(-)	(-)			(-)	+					[c]
NIH/3T3	c-H-*ras* mutant	Azatyrosine	+	+						+				
BALB/c	v-*mos*	Interferon	(-)				+		+			+	+	
Rat-1	v-*fos*	EMS, Rhodamine	+					+	+				+	−[d]
208F	v-*fos*	Adhesion	+										+	
Rat6	c-H-*ras* mutant inducible	Resistance to TPA/ZnSO$_4$	+	+									(+)	
F2408	v-K-*ras*, ts	UV Irradiation	+	+					+		+		+	−[e]
CCL 64	v-*fes*	MNNG rhodamine	+		+		+							

+, Cells resistant to transformation by indicated oncogene; −, cells sensitive to transformation; (+), heterogeneous response to oncogene expression; (−), revertants partially retransformed. AC, 5'-Azacytidine; EMS, ethane methosulfonate; CHP, *cis*-4-hydroxy-L-proline; AzaG, 8-azaguanine; BrdU, bromodeoxyuridine; TPA, 12-O-tetradecanoylphorbol 13-acetate; ts, temperature-sensitive mutant; MNNG, *N*-methyl-*N'*-nitro-*N*-nitrosoguanidine.

[a] Cells resistant to SV40 large T.
[b] Cells partially transformed by v-*myc*.
[c] Cells sensitive toward *hst*/*ret*.
[d] Cells sensitive to *trk*.
[e] Cells sensitive to *trk*, human papilloma virus, adenovirus type 12.
References see text.

rigenic. The oncogene resistance pattern of these CHP-derived revertants was compared with those from revertants derived from ouabain selection (Noda et al. 1983). Oncogenes were divided into four groups depending on their transforming activity when expressed in revertants or on their inability to confer the neoplastic phenotype. (Yanagihara et al. 1990). Oncogenes of the *ras* family form the first group in that they are suppressed by both ouabain-selected and CHP-selected revertants. Oncogenes with tyrosine kinase activity (*fes*, *src*, *fgr*) were suppressed by ouabain-selected revertants, but were able to retransform CHP-selected revertants. Group three includes oncogenes whose transforming activity is inhibited in CHP revertants, but not in ouabain revertants. Members of this group are oncogenes encoding serine-threonine kinases (*mos*, *raf*) and the nuclear oncogene *fos*. The oncogenes *sis* and *fms* were able to retransform both ouabain and CHP revertants and form group four. These results imply that many oncogene products exert their transforming activity via identical biochemical pathways. Thus, the number of different transformation pathways may be limited.

Two serum- and anchorage-dependent revertant cell lines were isolated from Ha-*ras* transfected NIH/3T3 cells by incubation in serum-free medium with bromodeoxyuridine. The revertants retained their nontransformed phenotype even if the level of mutated p21^{ras} was elevated by the introduction of additional copies of the oncogene (Yamada et a. 1990). Somatic cell hybrids generated by fusion of each of the revertant cell lines with nontransformed NIH/3T3 cells expressed the transformed phenotype. Fusion of the two revertant cell lines with each other resulted in a hybrid culture showing decreased anchorage requirements but only a slightly decreased serum dependence. The authors concluded that the revertants are recessive and have sustained mutations in effector genes necessary for the expression of the neoplastic phenotype. The significant reduction of anchorage dependence in the revertant x revertant fusion suggested that there may be two different complementation groups. Retransformation experiments with four types of oncogenes showed that the revertants and NIH/3T3 cells had an equal sensitivity toward c-*sis*, c-*neu*, and polyoma middle T oncogenes as indicated by colony growth in semisolid agar medium. In contrast, one revertant clone was relatively insensitive toward transformation by the v-*abl* oncogene.

The cell line B812, a derivative of Fisher rat fibroblasts, is a temperature-dependent cell mutant that shows a diminished frequency of foci after infection with different retroviruses (Inoue et al. 1983). For example, upon infection with Kirsten murine sarcoma virus the number of foci decreased more than 25-fold at the nonpermissive temperature of 39°C compared to the parental fibroblast line. Despite this phenotypic difference, the cells expressed nearly the same level of the *ras* transcript and p21^{ras} protein at the permissive and nonpermissive temperature. The ability to restrict focus formation was completely suppressed, when B812 cells were fused with wild-type parental cells. The temperature dependence was found after transformation with the oncogenes *mos*, *fos*, *src*, and *erbB2* but not after transformation mediated by polyomavirus middle-T antigen, adenovirus type 12, and human papillomavirus 16. These results suggested that the temperature-dependent transformation of B812 cells resulted from a mutation in a cellular factor(s) that is a common component of the transformation pathway used by *ras*, *mos*, *fos*, *src*, and *erbB2* oncogenes (Kizaka and Hakura 1989).

Revertants were obtained from transformed cells that expressed oncogenes other than *ras*. These cells exhibited resistance toward retransformation by *ras*, suggesting that some oncogenes exert their transforming activity via identical biochemical pathways. Morphological revertants of FBJ murine sarcoma virus-transformed Rat-1 cells were isolated after mutagenesis using a selection procedure based on the prolonged retention of the dye rhodamin-123 in mitochondria of transformed versus normal cells (Zarbl et al. 1987). The revertants continued to express the v-*fos* oncogene but exhibited features of nontransformed cells including loss of anchorage-independent growth and tumorigenicity. The revertants were resistant to retransformation by v-*gag-fos-fox*, v-Ha-*ras*, v-*abl*, and v-*mos* oncogenes but could be retransformed by the *trk* oncogene and polyoma virus middle T antigen. Somatic cell hybridization studies showed that the revertant phenotype was recessive, suggesting that the revertant cells have sustained mutations in cellular genes (or a single cellular gene) essential for neoplastic transformation. Following mutagenesis of rat fibroblasts transformed by Finkel-Biskis (FBR) murine sarcoma virus, revertants were isolated on the basis of their adherence to plastic culture dishes in the absence of divalent cations. These revertants were flat, contact-inhibited when grown to confluence, anchorage-dependent, and nontumorigenic in nude mice. Somatic cell hybrids of these revertants and the v-*fos*-transformed parental cells were nontransformed, suggesting the existence of a transformation-suppressing gene. These revertants were resistant to transformation by v-*fos*, c-*jun*, and activated c-Ha-*ras* genes (Wisdom and Verma 1990).

A revertant cell line of mink cells transformed by the Gardner-Arnstein strain of feline sarcoma virus showed a decrease in its proliferative capacity and saturation density, and a complete loss of its ability to form colonies in semisolid agar medium. The cell line was, however, tumorigenic in nude mice. Feline sarcoma provirus could be rescued from this cell line, and high levels of the v-*fes* oncogene were expressed. The revertant line was refractory toward transformation by retroviruses carrying v-*fes*, v-*fms*, and v-*ras* oncogenes. The transformation block was recessive, as fusion of the revertant line with the parental line yielded transformed hybrid cells. These results suggested that a defect in or limiting quantity of a gene product which functions downstream of the kinase and *ras* proteins abolished transformation in vitro (Haynes and Downing 1988).

4.2 Revertants Obtained After Treatment with Cytokines, TPA, Inducers of Differentiation, and Antibiotics

Revertants which maintained a nontransformed phenotype were obtained from NIH/3T3 cells transformed by a LTR-activated Ha-*ras* proto-oncogene by prolonged treatment with IFN-α/β (Samid et al. 1987). The nontransformed phenotype was maintained even after the cytokine treatment was discontinued (persistent reversion). Moreover, the cells continued to express large amounts of p21^{ras} and were resistant to transformation by transforming retroviruses harboring v-Ha-*ras*, v-Ki-*ras*, v-*abl*, or v-*fes* oncogenes. Retransformation was achieved by treating persistent IFN revertants with cytidine analogs. The authors suggested that DNA methylation is involved in IFN-triggered reversion. A revertant cell line was isolated from Moloney sarcoma virus transformed BALB/c cells after long-term IFN treatment (Gerfaux et al. 1990). The cells were not retransformable with viruses carrying v-*fes*, v-*fms*, and v-*fos* oncogenes. Infection with Kirsten sarcoma virus (v-Ki-*ras*) did not alter the flat morphology of revertant cells in vitro, but resulted in tumorigenicity. Persistent reversion was also achieved by treatment with IFN-γ of NIH/3T3 fibroblasts transformed by a mutant human Ha-*ras* oncogene (Seliger et al. 1991). Long-term treatment with the cytokine was not required. Rather, the population of cytokine-treated cells was subjected to the ouabain selection procedure described by Noda et al. (1983). The revertants resisted the transforming activity of v-Ha-*ras* and v-*mos* oncogenes but were partially retransformed by treatment with 5-azacytidine or by infection with recombinant retroviruses carrying the v-*abl*, v-*fes*, v-*myc*, or v-*src* oncogenes.

Krauss et al. (1992) took advantage of a rat cell line that stably overproduces protein kinase Cβ1 and harbors an activated c-Ha-*ras* oncogene controlled by an inducible metallothionein I promoter (MT I). Treatment of cells with TPA and ZnSO$_4$ led to cell death. However, surviving colonies arose in cell cultures either spontaneously or after induction by the mutagen ethyl methane sulfonate. Two variant cell lines, resistant to the cytocidal effects of TPA and ZnSO$_4$, showed a flat phenotype and did not grow in soft agar. The revertants continued to express PKCβ1 and retained an intact *ras* oncogene. These revertants were resistant to neoplastic transformation by v-*src* and v-*raf*. One of the two cell lines was refractory toward v-*fos*-induced transformation as well. When the two revertant cell lines were treated with TPA or infected with a retrovirus harboring an LTR-controlled Ha-*ras* oncogene, they regained a morphologically transformed phenotype but retained a strict anchorage dependence. Thus, *ras*- and PKC-mediated events leading to morphological transformation can be dissociated from those events involved in anchorage-independent growth. Moreover, the revertants failed to express the inducible exogenous *ras* oncogene as well as endogenous MT I and MT II genes on stimulation by the inducer. The authors proposed the existence of a *trans*-acting factor that represses the activity of genes controlling anchorage-independent growth. These unknown genes might have MT-like regulatory elements (Krauss et al. 1992).

Following treatment with the maturational agent sodium butyrate, the poorly differentiated human colon carcinoma cell line MIP-101 reverted to a more normal phenotype in that its population doubling time increased, the capacity to form colonies in semisolid agar medium was eliminated, and tumorigenicity was reduced (Niles et al. 1988). MIP-101 cells harbor an activated N-*ras* oncogene. The expression of the reverted phenotype was not correlated with decreased transcription of the oncogene (Stoddart et al. 1989). However, there was a marked reduction in the transforming activity of the genomic DNA from butyrate-treated MIP-101 cells. The molecular basis of this loss of transforming potential is not understood; however, the phenomenon appeared to be specific to MIP-101 cells. When a NIH/3T3 transformant harboring the human N-*ras* gene was treated with butyrate, similar phenotypic effects were observed. In contrast to MIP-101 cells, the transforming activity of genomic DNA from this cell line was unimpaired after butyrate-treatment. Flat revertants of v-*ras*-transformed rat kidney cells (KNRK) were generated to high efficiencies with sodium butyrate. The revertants expressed elevated levels of p21^{ras} (Ryan and Higgins 1989; Higgins and Ryan 1989).

The antibiotic azatyrosine selectively inhibited the growth of NIH/3T3 cells transformed by the activated human c-Ha-*ras* gene, while the growth of untransformed cells was unaffected (Shindo-Okada et al. 1989). More than 85% of azatyrosine-treated *ras*-transformed cells converted permanently to revertant cells that exhibited a flat morphology, contact inhibition, anchorage dependence, and a reduction of tumorigenicity in nude mice. The expression of the *ras* oncogene was unaffected. A similar reversion was caused by azatyrosine treatment of NIH/3T3 cells transformed by activated c-Ki-*ras*, N-*ras* and c-*raf* genes and of human pancreatic adenocarcinoma cells (PSN-1) harboring amplified Ki-*ras* and *myc* genes. In contrast, cells transformed by *hst* and *ret* oncogenes were refractory toward azatyrosine treatment. A summary of the pattern of resistance and sensitivity toward oncogene-mediated transformation observed in phenotypic revertants is shown in Table 1. The mode of azatyrosine action is not understood. It is very likely that azatyrosine affects the expression of effector genes necessary for signal transduction triggered by *ras* and other oncoproteins (Shindo-Okada 1989). Azatyrosine-induced revertant cell lines were also isolated from v-Ha-*ras* transformed human mammary epithelial cells (Kyprianou and Taylor-Papadimitriou 1992). Those cells were phenotypically nontransformed and nontumorigenic in nude mice, even though they contained prereversion levels of *ras* mRNA and p21 protein. Similarly, the expression of $p21^{ras}$ was not affected in revertants of *ras/myc* transformed BALB/c mouse embryo cells cultured in serum-free medium (Nomura et al. 1992).

Nalidixic acid (NaI) is an antibacterial agent that inhibits bacterial DNA gyrase, eukaryotic topoisomerase II, and viral replication in cultured cells. NaI reversibly suppressed growth in soft agar and reduced saturation density of BALB/3T3 cells transformed by methylcholantrene or of NIH/3T3 cells transformed by an activated c-Ha-*ras* oncogene (Kaneko and Horikoshi 1989). The amount of $p21^{ras}$ detected in NaI-treated cells was indistinguishible from control transformed cells. Two related compounds, oxolinic acid and pipemidic acid, were less effective than NaI in suppressing transformed cells. The authors speculate that these compounds indirectly modulate gene expression by interfering with topoisomerase activity.

Rat kidney cells infected with ts371 Kirsten murine sarcoma virus expressed the normal phenotype at the nonpermissive temperature (39.5°C) and the transformed phenotype at the permissive temperature (34°C; Itoh et al. 1989). Normal cells were characterized by expression of an unstable and less palmitylated $p21^{Ki-ras}$ and low GTP levels. When cells were treated with oxanosine, a guanine analogue antibiotic, at the permissive temperature, the transformed phenotype was blocked. GTP levels and $p21^{Ki-ras}$ stability resembled that of cells grown at the nonpermissive temperature.

A transformation-suppressed cell line R35, which accumulated a 23-kDa precursor of $p21^{ras}$, was isolated from *ras*-transformed NIH/3T3 cells. The revertant cell line was retransformed upon fusion with nontransformed Rat-2 cells. The defect that inhibited the expression of neoplastic transformation in R35 cells was due to a slower rate of palmitylation and polyisoprenylation of *ras* proteins (Huang and Axelrod 1991). Transformation by *ras* oncogenes can also be blocked at the posttranslational level by interfering with farnesylation that is necessary for membrane attachment of $p21^{ras}$ (for review see Schafer et al. 1989; Schafer and Rine 1992; Marshall 1993). Farnesyl synthesis and $p21^{ras}$ farnesylation can be blocked by lovastatin, an inhibitor of hydroxymethylglutaryl coenzyme A reductase. Lovastatin was shown to reverse abnormal morphology and loss of anchorage dependence as well as to reduce tumorigenicity in v-Ha-*ras* transformed rat liver epithelial cells (WB-*ras*; Ruch et al. 1993). In addition, lovastatin enhanced gap-junctional intercellular communication (GJIC) in WB-*ras* cells, suggesting that reduced GJIC plays a role in the expression of the transformed state (for review see Yamasaki 1991).

4.3 Reversion Induced by Inhibition of *ras* Expression

Benzamide (BA) is a NAD^+ site inhibitor of poly(ADP-ribose) polymerase which modifies numerous proteins by poly(ADP) ribosylation. This post-translational protein modification is involved in DNA repair, DNA replication, sister chromatid exchange, and cell differentiation. Probably, perturbed areas of chromatin generated around the regions of integration of exogenous genes are removed, when poly(ADP-ribose) polymerase inhibitors are active. BA induced a flat morphological phenotype of NIH/3T3 cells transformed by insertion of activated Ha-*ras*, Ki-*ras*, or N-*ras* genes, as well as truncated c-*raf* and *ret*-II oncogenes. The exogenous oncogene sequences were lost concomitant with the morphological reversion (Nakayasu et al. 1988). In contrast to this, endogenous *ras* sequences were not affected by BA treatment. Other inhibitors of poly(ADP ribose) polymerase were also able to induce the flat morphological phenotype. One of those, 1,2-benzopyrone (coumarin), prevented tumorigenesis of *ras*-transformed fibroblasts (Tseng et al. 1987). In contrast to benzamide-treated transformed cells, $p21^{ras}$ expression was not impaired. Thus, the action of the drug does neither directly nor indirectly involve the removal of the exogenous oncogene.

Melittin, an amphipathic peptide of 26 amino acids isolated from bee venom was shown to specifically select against cultured cells that express high levels of the *ras* oncogene (Sharma 1992). Cell killing by mellitin is specific for *ras* transformed cells, since nontransformed cells and cells transformed by the tyrosine kinase oncogene *ros*-1 or SV40 large T antigen were less sensitive. Acquisition of resistance to increasing concentrations of melittin is correlated with decreasing *ras* expression. Melittin was shown to be a potent activator of cellular phospholipase A_2 (PLA_2) in fibroblasts (Shier et al. 1979; Argiolas and Pisano 1983). Therefore, the authors suggest that hyperactivation of PLA_2 by melittin leads to the selective destruction of *ras*-transformed cells and that PLA_2 is an effector in the *ras* signal transduction pathway.

Masumoto et al. (1992) have screened microbial cultures for antitumor agents with "detransforming" activity. They identified a novel compound designated depudecin that inhibited the growth and induced the flat phenotype of NIH/3T3 cells doubly transformed by *ras* and *src* oncogenes. Morphological reversion was reversible. The effect of this drug on the expression of other transformed phenotypes was not reported.

An alternative strategy to diminish the concentration of the *ras* gene product is based on the introduction of antisense mRNA, hybridizing to mRNA encoding $p21^{ras}$ or a target molecule involved in *ras* signaling, of ribozymes, and antisense oligonucleotides. EJ *ras*-transformed NIH/3T3 cells were transfected with a plasmid containing an 84-bp sequence of the 5' end of the murine c-*fos* gene controlled by the dexamethasone-inducible MMTV promoter. The partial c-*fos* sequence was in antisense orientation relative to the promoter. Expression of the antisense RNA caused a significant decrease in the amount of the c-*fos* protein expressed after serum stimulation. The *ras*-transformed cells expressing antisense-*fos* RNA continued to overexpress *ras* and retained their proliferative capacity. However, their transformed properties were partially suppressed. Thus, inhibition of c-*fos* expression prevented the transforming ability of the *ras* oncogene. It was concluded that the *fos* proto-oncogene participates in *ras*-regulated signal transduction pathways (Ledwith et al. 1990).

Ras oncogene expression was successfully reduced by introduction of hammerhead ribozymes, designed to cleave c-Ha-*ras* mRNA mutated at codon 12, into NIH/3T3 cells transformed with the same oncogene. These ribozymes inhibited focus formation by about 50% and permitted to isolate morphological revertants of *ras*-transformed cells (Koizumi et al. 1992). Mutant Ha-*ras* mRNA expression was also selectively inhibited by antisense oligonucleotides centered around the point mutation at codon 12 (Monia et al. 1992) or codon 61 (Chang et al. 1991).

5 Gene Expression Associated with the Reversion of *ras*-Transformed Phenotypes

Various groups have identified genes that are specifically expressed in phenotypic revertants (Table 2), while expression is repressed in *ras*-transformed cells. Those genes were detected by differential hybridization techniques such that cDNA libraries were constructed with mRNA from revertants and that duplicate recombinant phage plaque lifts were hybridized with radioactively labeled cDNA from either transformed precursor cells or from revertants. The insert cDNAs of differentially hybridizing clones were then used as probes on northern blots of RNAs from normal, *ras*-transformed, and revertant cell lines (Müllauer et al. 1991; Contente et al. 1990; Krzyzoziak et al. 1992). An enrichment of revertant-specific cDNA clones was achieved by subtraction cloning. Single-stranded revertant cDNA was hybridized with an excess of mRNA from *ras*-transformed cells (Hajnal et al. 1993a,b; 1994). Alternatively, the overall protein synthesis in *ras*- or *fos*-transformed cells and revertants was compared by two-dimensional gel electrophoresis (Hoemann and Zarbl 1990; Higgins and Ryan 1991).

Elevated mRNA expression of α2 (type I) collagen (Müllauer et al. 1991), gelsolin, and α-actin (Müllauer et al. 1990) was found in the revertant cell line R1 (Kuzumaki et al. 1989). Revertant cell lines derived from transformed Rat-1 cells were resistant toward transformation by v-*fos* or *ras*

Table 2. Gene expression associated with phenotypic reversion of *ras* transformation

Gene/protein	Localization of gene product
α1 (Type I) collagen	Extracellular matrix
α2 (Type I) collagen	Extracellular matrix
Type III collagen	Extracellular matrix
Fibronectin	Extracellular matrix
Lysyl oxidase (*rrg*)	Cell supernate, extracellular matrix
Plasminogen activator inhibitor I	Extracellular matrix
α-Actin	Cytoskeleton
Gelsolin	Cytoskeleton
p92-5.7 (gelsolin variant)	Cytoskeleton
Vinculin	Cytoskeleton
Tropomyosin	Cytoskeleton
Cytochrome *b*	Mitochondria
Cytochrome *c* oxidase subunit II	Mitochondria
NADH dehydrogenase 1	Mitochondria
NADH dehydrogenase 4	Mitochondria
Intracisternal A particles	Cytoplasm
RhoB	Associated with cytoskeleton
H-rev 107	Membrane fraction, cytoplasm

oncogenes and resumed the synthesis of α1(type I) and α2(I) collagen. Moreover, it was shown that α1(I) procollagen but not α2(I) procollagen was regulated at the level of transcription in Rat-1 fibroblasts, transformed cells and revertants (Hoemann and Zarbl 1990). Gelsolin is an actin binding protein. A variant form of gelsolin, p92-5.7, was identified as a newly expressed protein in the revertant cell line R1 by two-dimensional gel electrophoresis (Fujita et al. 1990). This protein was not detected in a number of normal cell lines and primary embryo fibroblasts. Flat revertants derived from *ras*-transformed NIH/3T3 cells by azatyrosine treatment specifically expressed collagen type III. In addition, fibronectin expression that was downregulated on *ras* transformation was restored in revertants (Itoh et al. 1989; Krzyzoziak et al. 1992). Genes encoding (type I) procollagens were abundantly represented in a subtracted cDNA libary specific for normal rat 208F fibroblasts prior to the introduction of a Ha-*ras* oncogene. Reexpression of different collagen genes was found in phenotypic revertants (Hajnal et al. 1993a). Elevated expression of cytoskeletal and extracellular matrix components in revertants was confirmed at the protein level. The expression of nonmuscle tropomyosin was restored to pretransformation levels in ouabain-selected revertants of v-Ki-*ras* transformed NIH/3T3 cells (Bassin and Noda 1987). Plasminogen-activator inhibitor type I (PAI-1), an extracellular matrix-associated 52-kDa protein, was induced during the generation of the revertant phenotype. p52(PAI-1) expression was regulated at the level of mRNA abundance. A three- to fourfold increase in cytoskeletal deposition of gelsolin and vinculin (Higgins and Ryan 1991) was reported in butyrate-induced revertants of v-*ras* transformed rat kidney cells (Higgins and Ryan 1989; Ryan and Higgins 1989). This pattern of microfilament-associated proteins correlated with the revertant-specific reorganization of microfilament and focal contact formation. A similar increase in actin content was observed, whereas the abundance of intermediate filament components vimentin and lamins remained unaltered. These results suggest a role for both cytoskeletal proteins and extracellular matrix components in restructuring the normal cytoarchitecture during the process of reversion.

Contente et al. (1990) described a gene, designated *rrg* (*ras* recision gene) in the revertant cell line PR4, a persistent revertant of LTR-c-Ha-*ras* transformed NIH/3T3 cells treated with IFN-α/β. The expression of *rrg* was reduced to an undetectable level in *ras* transformed cells and incompletely restored in the revertant. Functional studies provided circumstantial evidence that *rrg* expression was sufficient to restore a more normal phenotype in *ras*-transformed NIH/3T3 cells. When persistent revertants were stably transfected with a *rrg* cDNA antisense expression vector, *rrg* mRNA was reduced in association with retransformation. Although this finding qualified *rrg* as a potential transformation suppressor gene, it is doubtful whether the inactivation of *rrg* is a necessary step for the tumorigenic conversion of a normal cell. Expression of antisense rrg mRNA in NIH/3T3 cells did not cause neoplastic transformation. The downregulation of *rrg* expression was not specific for *ras*-transformed cells. A significant decrease in *rrg* expression was also observed in v-*raf*, and v-*fes* transformed cell lines. Variable amounts of *rrg* mRNA were detected in cells harboring a SV40 early promoter–c-*myc* recombinant gene or v-*abl*, v-*sis*, and v-*mos* oncogenes.

Sequence comparisons between *rrg* cDNA and a 2672 bp cDNA of rat aorta lysyl oxidase (Trackman et al. 1990) demonstrated that *rrg* encodes lysyl oxidase (Kenyon et al. 1991). Thus, the product of *rrg* is a copper-dependent amine oxidase that catalyzes the oxidative deamination of peptidyl lysine residues in procollagen and proelastin (Trackman et al. 1990). A nonenzymatic condensation reaction forms lysine-derived cross-links in the mature extracellular matrix proteins. More than 90% of lysyl oxidase activity was found in the cell supernatant. A significant reduction of enzyme activity correlated well with the tumorigenic phenotype of *ras*-transformed NIH/3T3 cells and of rerevertants induced by antisense *rrg* expression

(Kenyon et al. 1991). Upregulation of lysyl oxidase was found in other revertant cells as well. These include revertants obtained from *ras*-transformed NIH/3T3 cells by treatment with azatyrosine (Kryzoziak et al. 1992) and IFN-γ (H. Oberhuber, R. Schäfer, B. Seliger, unpublished) as well as spontaneous morphological revertants of *ras*-transformed rat 208F cells (Hajnal et al. 1993b). The posttranslational modification of extracellular matrix precursor polypeptides by lysyl oxidase may modulate cell morphology and interactions between cells and extracellular matrix. It has been speculated that lysyl oxidase may indirectly modulate important signal transduction pathways by oxidation of membrane-bound receptors (Kenyon et al. 1991).

Four different mitochondrial genes encoding cytochrome *b*, cytochrome *c* oxidase subunit II, as well as NADH dehydrogenases 1 and 4 were specifically expressed in the flat revertant cell line R1 (Müllauer et al. 1991). Azatyrosine-induced revertants derived from *ras*-transformed human mammary epithelial cells showed an increased level of K-*rev*-1 mRNA expression (Kyprianou and Taylor-Papadimitriou 1992). Therefore, the reversion of transformed human mammary epithelial cells may be mediated by an elevated level of K-*rev*-1 expression (see next section). Azatyrosine-induced revertants expressed *rhoB* and sequences corresponding to that of murine retrovirus-like intracisternal particles (IAP). IAP-related cDNAs represented about 50% of the clones specific for the reverted state and identified by differential cDNA hybridization (Kryzosiak et al. 1992). Rho, a *ras*-related GTP binding protein, rapidly stimulated stress fiber and focal adhesion formation after microinjection into serum-starved Swiss 3T3 cells (Paterson 1990; Ridley and Hall 1992). These findings suggest that Rho is an essential protein for the coordinated assembly of focal adhesions and actin stress fibers in growth-stimulated cells. Dysfunction of *rho* by downregulation of its expression may at least partially be responsible for the deformation of stress fibers on oncogene-mediated transformation. The reestablishment of *rho* expression may significantly contribute to those events that mediate the normal morphology of revertants.

By subtraction cloning Hajnal et al. (1994) have isolated a novel gene, designated H-*rev*107, which is specifically expressed in a revertant clone obtained from Ha-*ras* transformed rat 208F cells. Rat cells transformed by Ha-*ras* or v-*src* oncogenes do not contain detectable amounts of H-*rev107* mRNA, whereas logarithmically growing fibroblastic cell lines such as NIH/3T3 and 208F express very little. In contrast to this, H-*rev107* mRNA was abundant in *ras*-transformation-resistant cell lines REF-52 and EK-3. Introduction of the adenovirus E1A oncogene into REF-52 cells, which abolishes the resistance toward *ras* transforming activity (Franza et al. 1986), repressed the H-*rev*107 gene. The function of H-*rev*107 is still unknown. The protein product of the H-*rev*107 gene exhibits an apparent molecular mass of 18,000 and has no sequence similarity to known proteins. P18$^{H\text{-}rev107}$ has a hydrophobic C-terminal stretch of amino acids. Two forms of the protein were distinguishible in cellular extracts of REF-52 cells by SDS gel electrophoresis. The faster migrating form was localized in the cytoplasm, the other was present predominantly in the membrane fraction. When normal rat 208F cells were subjected to density-dependent growth arrest, membrane-associated p18$^{H\text{-}rev107}$ was accumulated. Growth arrest induced by serum starvation did not induce expression of H-*rev*107 (Hajnal et al. 1994).

6 *ras* Transformation Suppressor Genes

In the previous section several genes were discussed that appear to be involved in the maintenance of the reverted state in derivatives of *ras*-transformed cells (Table 2). It is desirable to know which genes have the capability to induce phenotypic reversion. Functional assays have been instrumental in identifying transforming genes. A similar approach is difficult to design for genes whose function is to constrain growth since their expression does not confer a growth advantage on cells. Following transfection of genomic DNA or of complementary DNA from normal cells into *ras*-transformed recipient cells, revertant clones were obtained by drug selection and/or morphological selection. DNA sequences were identified capable of counteracting *ras*-mediated transformation (Schäfer et al. 1988; Noda et al. 1989; Noda 1990; Cutler et al. 1992). Genomic DNA prepared from human placental cells was transfected into rat FE-8 cells, a neoplastically transformed derivative of 208F cells expressing an activated human Ha-*ras* oncogene (Schäfer et al. 1988). The population of transfected FE-8 cells, representing the entire donor genome in about 6 000 transfectant clones, was subjected to selection with ouabain. This drug was used to selectively eliminate those transfectants that continued to express the neoplastic phenotype. A few revertant clones survived ouabain-selection and continued to express the reverted phenotype. In a subsequent transfection cycle, the reverted phenotype was passaged onto FE-8 cells. A recombinant DNA library was constructed from the DNA of a secondary transfectant. Screening this library with a repetitive human Alu probe identified one recombinant clone that carried the suppressing activity. However, the mechanism of suppression is still obscure, since sequence analysis of cloned transfected human DNA revealed an extensive structural DNA rearrangement and no open reading frame. Possibly, the suppressing activity is mediated by transcription of DNA sequences harboring a protein binding site (M. Kiess and

Table 3. Suppressors of *ras* oncogene induced transformation

Gene	Subcellular localization and biochemical properties of gene product	Proposed cellular function
*Krev-1/rap1A/*smg 21	Membrane, GTP-binding, GTP-hydrolysis	Signal transduction oxygen radical production
rsp-1	?	Signal transduction
GAP	Cytoplasmic[a], activation of GTP-hydrolysis	Signal transduction
NF-1 (neurofibromin)	Membrane (cytoplasmic)[b], activation of GTP-hydrolysis	Signal transduction
H-*ras*	Membrane, GTP-binding, GTP-hydrolysis	Signal transduction
Thy-1	Cell surface, glycoprotein	Cell-cell recognition
gas-1	Transmembrane protein	Cell cycle control
p53	Nuclear phosphoprotein	Transcriptional regulation, growth control

[a] May translocate to membrane.
[b] Membrane-associated in many tissues, cytoplasmic in brain.

R. Schäfer, unpublished observations). The two other groups have used cDNA expression libraries for the transfer of suppressing activity into *ras*-transformed cells and been able to identify two functionally active genes, designated *Krev*-1 (Kitayama et al. 1989) and *rsp*-1 (Cutler et al. 1992). A list of genes capable of suppressing transformation by mutated Ras is shown in Table 3.

6.1 *Krev-1/rap1A*

6.1.1 *Identification by Functional Assay*

Seven flat revertant cell lines were isolated from DT, a Kirsten murine sarcoma virus transformed NIH 3T3 cell line following transfection of a human fibroblast cDNA library (Noda et al. 1989; Kitayama et al. 1989). The human foreskin fibroblast expression library was constructed in the eukaryotic expression vector pcD2 harboring the *neo* gene as a selectable marker (Chen and Okayama 1987). The transfected cDNA was recovered from one revertant clone, R16, in the following way: *Sal*I-digested DNA from R16 cells was circularized by ligation and transformed into competent *Escherichia coli*. Although the *neo* gene is controlled by a eukaryotic promoter in pcD2, it is able to confer weak kanamycin resistance on bacteria. Several kanamyin-resistant plasmid clones were transfected into DT cells. One clone, pKrev-1, conferred the revertant phenotype onto v-Ki-*ras* transformed cells (Kitayama et al. 1989). Morphological reversion of DT cells requires high *Krev* -1 expression. *Krev*-1 cDNA encodes a 184 amino acid open reading frame, corresponding to a protein with a calculated molecular mass of 21 kDa. This open reading frame shares strong structural similarity with *ras* proteins. The similarities are confined to the β-phosphoryl group and guanine binding domains of the Ha-*ras* protein. Moreover, the effector domain in the Ha-*ras* (amino acid residues 32–44) and the *Krev*-1 proteins is identical. As *ras* proteins, the predicted *Krev*-1 protein has a CAAX consensus sequence at its extreme carboxy-terminal end (C, cysteine residue; A, aliphatic amino acid; X, any amino acid residue). In the Ha-*ras* protein, this amino acid sequence is necessary for membrane attachment and transforming activity. Protein domains with less similarity between *Krev*-1 and *ras* are the residues 59–63, known to harbor point mutations in oncogenically activated *ras* proteins, residues 72–85, found in high molecular weight G-proteins, and residues 152–164 that are essential for Ha-*ras* transforming activity. The α-helical region located at the carboxy-terminal end of *Krev*-1 is shorter than those of *ras* proteins. The *Krev*-1 gene is expressed in many tissues including brain, thymus, lung, spleen, colon, and ovary.

One year before the identification of *Krev*-1, three genes, designated *rap1A*, *rap1B*, and *rap2*, were isolated from a human cDNA library by low-stringency hybridization with the *Drosophila Dras*3 gene (Pizon et al. 1988a,b). In the same year a novel small GTP binding protein was purified from bovine brain crude membranes and its cDNA cloned from a bovine brain cDNA library. The protein was designated smg p21 (Kawata et al. 1988). While *rap1A* is identical to *Krev*-1, the smg p21 gene is identical to *rap1B*. The *rap1A* proteinwas independently isolated from human neutrophils (Bokoch et al. 1988; Quilliam et al. 1990). The product of the *Krev*-1/*rap1A* gene was copurified as a component of the superoxide generating system from human neutrophils (Quinn et al. 1989). The *rap1B* protein was also purified from human platelets (Ohmori et al. 1989; Matsui et al. 1990) and bovine aorta smooth muscle (Kawata et al. 1989; 1991). A close relative of the *rap2* gene, designated *rap2B*, was isolated from a human platelet cDNA library (Farrell et al. 1990; Ohmstede et al. 1990; Lerosey et al. 1991). The *rap*1 and *rap*2 proteins exhibit 70% identity at the amino acid level while *rap1A*/*rap1B* and *rap2A* and *rap2B* proteins are 95% and 90% identical, respectively (for review see Bokoch 1993).

The frequency of morphological reversion of Ki-*ras* transformed DT cells induced by overexpression of *Krev*-1 was rather low (2%–5% of total transfectants; Kitayama et al. 1989). Site-directed mutagenesis of the *Krev*-1 cDNA was performed to further investigate the biological activity of the *Krev*-1 gene. Point mutations in codon 38 resulting in the substitution of Asp by Ala or Asn inhibited the ability of the gene to cause phenotypic reversion. The reversion frequency was enhanced between two- and fivefold by amino acid exchanges such as Gly (codon 12) to Val and Gln (codon 63) to Glu (Kitayama et al. 1990). In addition to suppressing the transformed properties of DT cells, these *Krev*-1 mutants showed strong tumor suppressor activities in HT 1080 fibrosarcoma cells harboring the activated N-*ras* gene. In another study, chimeric *Krev*-1/Ha-*ras* genes were assayed for transforming or suppressing activity (Zhang et al. 1990). The critical sequences for transformation by v-Ha-*ras* or suppression by *Krev*-1 were localized to the NH_2-terminal 54 amino acids. Amino acids 32–44 are identical in *Krev*-1 and *ras* (Kitayama et al. 1989). The *ras*-specific and *Krev*-1 specific amino acids immediately surrounding these amino acid residues, particularly residues 26, 27, 30, 31, and 45 determined whether the protein has transforming activity or acts as a transformation suppressor (Zhang et al. 1990; Marshall et al. 1991; Nur-e-Kamal et al. 1992). This region of $p21^{ras}$ represents the effector binding domain (Barbacid 1987; Grand and Owen 1991), indicating that *Krev*-1 suppresses *ras*-mediated transformation by competing with the oncoprotein for a target molecule (or molecules).

The suppressing effect of *Krev*-1 was also assessed in another human tumor cell line. Cell clones expressing exogenous *Krev*-1 mRNA were derived from the anaplastic lung carcinoma cell line Calu-6. In contrast to parental cells, revertant lines had a flat or squamous morphology and proliferated continuously, independent of a low serum concentration in the medium. After subcutaneous inoculation into nude mice, the revertant cells formed tumors albeit with a reduced growth rate. In a tracheal transplantation assay, the invasive properties of parental cells and revertant lines were compared. However, a consistent loss of invasiveness was not found in different revertant cell lines. Only one revertant line was able to metastasize spontaneously, while parental cells exhibited a high metastatic potential (Zucker et al. 1992). In view of these weak effects on tumorigenicity and metastactic potential, it was of interest to find out whether *Krev*-1 may prevent neoplastic transformation in normal cells when expressed with an oncogene. Preneoplastic Rat-2 cells were cotransfected with a low amount of plasmid DNA, in which the polyomavirus early promoter controls middle T (mT) antigen expression, and with increasing amounts of p*Krev*-1 DNA (Jelinek and Hassell 1992). mT antigen-mediated morphological transformation of Rat-2 cells was inhibited in a dose-dependent manner. Inhibition was about 85% of the control at the highest dose of p*Krev*-1 DNA used. This inhibitory effect appeared to be specific, since morphological transformation by SV40 large T antigen or the frequency of puromycin-resistant colonies was unaffected by *Krev*-1 DNA cotransfer in control experiments. Moreover, the same authors isolated 16 morphological revertant cell lines after transfection of p*Krev*-1 into mT-antigen transformed cells. Morphological revertants represented only about 1% of the total transfectants, thus confirming the rather weak activity of the transformation suppressor described earlier (Kitayama et al. 1989). One exceptional clone had lost the oncogene. The other 15 clones differed from their transformed precursor cell line in their reduced growth rates and low capacity to form colonies in semisolid agar medium. The ability to form colonies in agar medium was inversely proportional to the amount of *Krev*-1 mRNA and protein expressed. The results of this study confirmed an earlier result (Smith et al. 1986) that p21^{ras}, which is counteracted by *Krev*-1, lies downstream of mt antigen and pp60 c-*src* in the same mitogenic signal transduction pathway (Jelinek and Hassell 1992).

Although Rap2 proteins contain an effector domain identical to Ras, they do not suppress Ras-mediated transformation (Schweighoffer et al. 1990; Jimenez et al. 1991). The intrinsic GTPase activity of Rap2A was not stimulated by rasGAP and there was no competition with Ras for interaction with *ras* GTPase activating protein (GAP; Lerosey et al. 1992).

If the *Krev*-1 gene plays an important role in the multistep process of tumorigenesis, one would expect to find tumors in which its structure or expression is altered. Therefore, the expression of *Krev*-1 was investigated in tumorigenic cells. *Krev*-1 related transcripts were abundant in colonic cells (Kitayama et al. 1989). The locus of *Krev*-1/*rap1A* has been mapped to human chromosome 1p12-p13. The *rap1B* and *rap2* genes have been assigned to bands 12q14 and 13q34 of the same chromosome, respectively (Rousseau-Merck et al. 1990). Both structural and numerical abberrations of the short arm of chromosome 1 are frequent in tumors, including colorectal tumors (Mitelman 1990; Muleris et al. 1990). Moreover, oncogenic activation of Ki-*ras* genes is frequent in colorectal cancer (Vogelstein et al. 1988). These findings prompted an investigation of the *Krev*-1/*rap1A* locus in human colon cancer. Young et al. (1992) identified a *Krev*-1-specific *Bcl*I restriction fragment length polymorphism (RFLP) that permitted to search for allelic deletions by comparison of the constitutional chromosome content and of tumor tissue. Only one out of 18 tumors from informative cancer patients showed allelic loss at this locus, suggesting that loss of heterozygosity at the *Krev*-1 locus is a rare event in colorectal tumors. Culine et al. (1989) have reported northern blot analysis of the three *rap* genes in 41 primary human tumors, including 13 lymphomas, 14 carcinomas, 6 sarcomas, various tumors of the nervous system, and germinal neoplasms of the testis. A significant decrease in the expression of the *rap1A* gene was observed in one salivary gland adenocarcinoma and in the sarcomas tested with the exception of the leiomyosarcoma. In contrast, *rap1B* was highly and uniformly transcribed in nearly all normal and malignant tissues analyzed. Usually, detection of *rap2* expression was only possible in polyadenylated RNA. The *rap2* expression was absent in sarcoma poly(A) RNA samples. The tumors were chosen on the basis of a low incidence of *ras* mutations. The significance of *rap* gene downregulation in human multistep carcinogenesis is not clear.

In an attempt to investigate the potential role of *rap1A* and *rap1B* genes in mammary tumorigenesis in rats, mammary tissues from tumor-susceptible Sprague-Dawley and tumor-resistant Copenhagen rats were screened for deregulation of *rap* gene transcription and for mutations. However, differences between those two inbred rat strains were not found (Hong et al. 1990; Hsu and Gould 1991).

6.1.2 *Mechanism of Interference with ras Function*

The greatest divergence in amino acid sequence between Ras and Rap1A occurs at the C-terminus. Krev-1/Rap1A is posttranslationally modified by a C_{20} geranylgeranyl group at Cys-181. The last three amino acids are proteolytically cleaved off and the terminal prenylated cysteine is carboxymethylated (Buss et al. 1991). Ras contains a C_{15} farnesyl at its terminal cysteine residue. As shown by functional analysis of Ras/Rap1A proteins, farnesyl and geranylgeranyl moieties are mutually interchangeable for transforming and suppressing activity, respectively (Buss et al. 1991; Cox et al. 1992). Fractionation of cell extracts has shown that Rap1 and Ras were not associated (Grand and Owen 1991; Beranger et al. 1991, Kim et al. 1990). In Rat-1 fibroblasts and Hep2 epidermoid carcinoma cells, Rap1 was associated with a Golgi-like structure, while Ras was found in the cytoplasmic membrane fraction (Beranger et al. 1991). Association of Rap1 with both the plasma membrane and granular structures was, however, found in neutrophils (Maridonneau-Parini and de Gunzburg 1992; Quinn et al. 1992). Rap1A was shown to be phosphorylated at Ser-180 by cyclic AMP-dependent protein kinase in vitro. (Kawata et al. 1989; Lapetina et al. 1989; Quilliam et al. 1991). $p21^{rap1A}$ appears to be identical to thrombolamban, a 22-kDa protein, which serves as a major substrate for protein kinase A in platelets (Fischer and White 1987). It has been speculated that Rap1A is involved in the release of Ca^{2+}, and that phosphorylation of this protein by c-AMP activated protein kinase C inhibits this function (Kawata et al. 1989; Lapetina et al. 1989). Phosphorylation upon differentiation was demonstrated in intact HL-60 cells (Quilliam et al. 1991). Phosphorylation of Rap1A did not affect guanine nucleotide binding and hydrolytic activity.

The identity of p21^{ras} and p21^{rap1A} proteins in the effector region suggested that GAP, a known effector of *ras* action (for review see Downward 1992; Hall 1992) also interacts with the product of *Krev*-1. Therefore the ability of GAP to interact with Rap1A-p21 was tested in vitro. Rap1A bound tightly to GAP and effectively inhibited GAP-mediated Ras-GTPase activity. The binding of GAP to Rap1A was GTP dependent. The affinity of Rap1A-GTP for rasGAP was more than 50-fold greater than that of Ras-GTP. Thus, Rap1A may compete with Ras for its downstream effector. The GTPase activity of Rap1A was not stimulated by rasGAP, suggesting the Rap1A/rasGAP complex was catalytically nonfunctional (Frech et al. 1990; Hata et al. 1990).

Rap1A does not only associate with rasGAP but also with a GTPase activating protein that is specific for Rap1 and shows no significant similarity to rasGAP (Rubinfeld et al. 1991, 1992). Initially, two smg p21 GAP proteins were isolated from bovine brain and shown to be distinct from p21*ras* GAP and p20^{rho} GAP (Kikuchi et al. 1989). Later, a membrane-associated rapGAP protein was purified from differentiated promyelocytic HL60 cells (Polakis et al. 1991). The gene encoding rapGAP was isolated from a human brain cDNA library (Rubinfeld et al. 1991). Apart from the sequence difference, rap1GAP and rasGAP can be distinguished by their mode of interaction with the GTP binding proteins Ras and Rap1: GTP hydrolysis is brought about by different mechanisms. Glu-61 is essential for GTP hydrolysis by Ras. The same amino acid residue is not critical for this reaction by Rap1. When Thr-61 of p21^{rap1A} was mutated to glutamine, rasGAP could accelerate the rate of GTP hydrolysis (Hart and Marshall 1990). Binding of Rap1 to its specific GAP occurs with both GTP-bound and GDP-bound forms, while Rap1A binding to rasGAP is GTP-dependent. It was proposed that the function of Rap1 depends upon its interaction with both rasGAP and rap1GAP (Rubinfeld et al. 1991). First, binding of Rap1 to rasGAP may inhibit the interaction of Ras and rasGAP (Frech et al. 1990; Hata et al. 1990). Second, upon binding of Rap1 to its specific GAP, bound GTP is hydrolyzed and the protein switched into its GDP-bound state. Since the binding of Rap1 to rasGAP is dependent upon GTP, the ability of Rap1 to antagonize Ras (i.e., to compete for rasGAP binding with p21*ras*) would be regulated by interaction with its own specific GAP. The binding of p21^{ras} to rasGAP is thought to result in a growth-promoting signal. When Rap1-GTP binds to the same downstream effector, this growth signal would be disrupted (for review see Hall 1990; McCormick 1989). It is assumed that Rap1 would be in the GDP-bound form in growing cells, due to its interaction with rap1GAP, and would be unable to perturb Ras function. It cannot be excluded however, that Rap1 signaling is transmitted via a separate pathway which also requires the GTP-bound form of the protein (Rubinfeld et al. 1991). The issue is further complicated by the fact that rasGAP may act as an upstream regulator of Ras rather than as a downstream target (Hall 1990).

A GDP/GTP dissociation stimulator, Rap1-GDS, has been identified and its cDNA been cloned (Yamamoto et al. 1990; Kaibuchi et al. 1991; Kikuchi et al. 1992). The 53-kDa protein is a relative of the yeast CDC25 and SCD25 proteins and is able to catalyze GDP/GTP exchange for Rap1A, Rap1B, Ki-Ras, Rho A, and Rac1 (Mizuno et al. 1991; Kotani et al. 1992). GTP binding of Rap1B may be regulated by binding to certain phospholipids via its C-terminal domain and by cyclic AMP dependent kinase (see Bokoch 1993).

6.1.3 Multiple Biological Activities

Rap1A was shown to modulate different physiological processes which involve signal transduction mediated by Ras. These processes include the short-term stimulation of DNA synthesis in Swiss 3T3 cells, transcription of the immediate early gene c-*fos*, opening of K^+ channels in atrial membranes, mitotic maturation in *Xenopus*, and photoreceptor development in *Drosophila*.

The cellular level $p21^{Krev-1/rap1}$ was elevated to about 120-to 150-fold of the endogenous level by microinjection into Swiss 3T3 cells of the protein synthesized in *E. coli* or *Spodoptera frugiperda*. Both $p21^{Krev-1}$ and $p21^{Ki-ras}$ (Val-12) induced DNA synthesis when in their GTP-bound form. Their action differed in that Ras induced membrane ruffling irrespective of the presence or absence of insulin, whereas the GTP-bound form of Rap1A required insulin. Ras microinjection also resulted in the decrease of stress fibers and rounding up of cells. In contrast, Rap1A did not cause morphological transformation (Yoshida et al. 1992).

The effect of *Krev-1/rap*1A cDNA on the c-*fos* promoter/enhancer linked to the luciferase reporter gene (c-*fos* luciferase) was examined by transient transfection into Cos cells (Sakoda et al. 1992). While the basal expression of the reporter gene in NIH/3T3 cells was rather low, c-*fos* luciferase expression was elevated in NIH/3T3 cells constitutively expressing a mutant Ki-*ras* (Val-12) or activated c-*raf*-1 kinase. Treatment with 12-tetradecanoyl phorbol 13-acetate (phorbol ester, TPA), platelet-derived growth factor (PDGF), or dibutyryl cyclic AMP (Bt2cAMP) also resulted in enhanced c-*fos* luciferase expression. Overexpression of *rap*1A and *rap*1B cDNAs inhibited the stimulation of c-*fos* luciferase mediated by $p21^{ras}$, PDGF, or TPA. However, c-raf-1 kinase or Bt2cAMP stimulation of the reporter gene was not affected. Thus, Rap proteins inhibited the signal transduction pathway from the PDGF receptor via protein kinase C and $p21^{ras}$ effectors to the c-*fos* promoter/enhancer, whereas signaling from *raf* kinase and cAMP-dependent protein kinase was not impaired.

Rap1A was able to inhibit the concerted negative effects of Ras and Ras-GAP on opening of M_2 muscarinic receptor-regulated K^+ channels of atrial membranes (Yatani et al. 1990, 1991). Particularly, an intact Rap1A effector domain was required for antagonistic function and Rap1A did not block the inhibition of channel opening by a form of GAP that was independent of Ras (Yatani et al. 1991).

Xenopus laevis oocytes (stage 6) can be induced to proceed through mitotic maturation, also called germinal vesicle breakdown (GVBD), by progesterone, insulin, or insulin-like growth factor 1 (IGF-1). GVBD is blocked after microinjection of antibodies against the *ras* gene product (Korn et al. 1987) and stimulated oncogenic Ha-*ras* proteins, such as the Val-12 mutant (Birchmeier et al. 1985). GVDB induced by Ras was effectively blocked by coinjection of equimolar amounts of the Rap1B. This inhibition of *ras*-induced GVBD was in turn lifted when IGF-1 was added to the incubation medium (Campa et al. 1991).

The human and *Drosophila rap* genes are extremely similar (Pizon et al. 1988). The Phe-157 residue in *Drosophila rap*1 corresponds to Phe-158 in human Rap1A and Phe-156 in human Ha-Ras. This amino acid residue is one of only five residues outside the GTP binding region which are conserved in every member of the *ras* superfamily. Substitution of Leu for Phe in *Drosophila Rap*1 causes a gain-of-function mutation resulting in the absence of an R7 photoreceptor cell in many omnatidia (*roughened* mutation; Hariharan et al. 1991). The correct diffentiation of R7 cells involves a signal transduction pathway, in which the product of the sevenless receptor kinase gene (Hafen et al. 1987) and the *ras*1 gene are important components (Simon et al. 1991). It is possible that Rap1 antagonizes Ras1 in the process of R7 diffentiation. Loss-of-function mutations of *Drosophila rap*1 were lethal (Hariharan et al. 1991).

Rap1A interacts with the phagocyte NADPH oxidase/cytochrome *b* system. Neutrophils and other phagocytic cells utilize a multicomponent enzyme complex for the reduction of molecular oxygen to form superoxide anion and other toxic oxygen metabolites in response to contact with opsinized micro-organisms ("respiratory burst"). This enzyme complex consists of at least four proteins including cytochrome b_{558} (see Bokoch 1993 for a review). The first indication that Rap1A might modulate this enzyme complex derived from a report by Quinn et al. (1989) who copurified a protein of relative molecular mass 22 000 with cytochrome *b*. The sequence of the cDNA was identical to the *Krev*-1 sequence. The possibility of a direct association of *Krev*-1/*rap*1 protein with cytochrome *b* was confirmed by immunoaffinity purification methods with matrices conjugated with anticytochrome and anti-Ras antibodies. The elution pattern suggested an association of immunopurified neutrophil cytochrome *b* and the *ras*-related protein. Complexes between the two proteins were also detected in vitro with purified cytochrome *b* and recombinant Rap1A expressed in baculovirus-infected Sf9 cells (Bokoch et al. 1991). The guanosine-5'-O-(3-thiotriphosphate; GTP-γS) bound form of Rap1A bound more tightly to cytochrome *b* than did the GDP-bound form. Complex formation between Rap1A and cytochrome *b* was inhibited by phosphorylation of Rap1A by protein kinase A (Bokoch et al. 1991). The significance of the interaction of Rap1A with NADPH oxidase-associated cytochrome *b* is not yet understood (Bokoch 1993).

6.2 *rsp*-1

An expression cloning assay, similar to the one described by Noda et al. (1989) was used to isolate a novel cDNA, referred to as *rsp*-1, which is able to suppress the v-*ras* transformed phenotype (Cutler et al. 1992). The expression library was constructed on the basis of RNA from the revertant cell line CHP9CJ (Yanagihara et al. 1990; see above). The v-*ras*-transformed NIH/3T3 cell line DT was used as recipient. After selection in G 418 and treatment of transfectants with the aminoglycoside ouabain (Noda et al. 1983), flat revertants were isolated. Several cDNAs isolated from one primary transfectant clone were able to confer the flat phenotype onto DT cells at a moderate frequency of 10%–20%. Full-length *rsp*-1 cDNA contains a 831-bp open reading frame encoding a 277 amino acid protein. The mouse and human *rsp*-1 genes are highly conserved. A major transcript of 1.7 kb was detected in NIH/3T3 cells, DT cells, and several oncogene transformants. The *rsp*-1 coding sequence was cloned into a retroviral vector. DT cells were infected with the recombinant vector and the ability to suppress soft agar colony formation was assessed. Overexpression of *rsp*-1 resulted in a 25%–75% reduction in colony formation. *Rsp*-1 cDNA controlled by an inducible promoter was able to revert the anchorage independent phenotype of Ki-MuSV-infected NOG8 mouse mammary epithelial cells. *Rsp*-1-infected NIH/3T3 cells grew more slowly than vector control cell lines suggesting a negative or toxic effect on the growth of NIH/3T3 cells. One *rsp*-1-infected cell line exhibited resistance toward retroviruses containing v-Ki-*ras* and v-Ha-*ras* oncogenes, but were transformed by v-*src* and v-*mos* oncogenes.

The predicted *rsp*-1 protein contains several leucine-based repeats homologous to similar repeats found in the regulatory region of yeast adenylate cyclase. In *S. cerevisiae* this region of the protein is necessary for regulation of the cyclase by Ras (Colicelli et al. 1990; Field et al. 1990; Suzuki et al. 1990). Since Ras does not regulate adenylate cyclase in higher eukaryotic cells, Cutler and colleagues (1992) have suggested that *rsp*-1 is not a component of the adenyl cyclase pathway in spite of its structural similarity to the yeast protein. Rather, they proposed that *rsp*-1 either interferes with a downstream target molecule in the Ras-signaling pathway or may represent a downstream target itself. The normal cellular function of *rsp*-1 is still unknown.

6.3 *GAP*

Xenopus oocytes contain an enzymatic activity which stimulates the rate of hydrolysis of GTP bound to microinjected $p21^{ras}$. The responsible enzyme was designated GTPase activating protein, or GAP (Trahey and McCormick 1987). The corresponding gene was cloned from bovine brain (Vogel et al. 1988) and human placenta (Trahey et al. 1988). The protein product of the GAP gene exhibits a molecular mass of 120 kDa. The sequence contains two SH2 (*src* homology) regions and one SH3 motif. SH2 domains mediate protein-protein interactions by binding to phosphotyrosine containing sequences (Cantley et al. 1991). Although $p120^{GAP}$ binds to wild-type and mutant $p21^{ras}$, it only stimulates GTP hydrolysis on normal $p21^{ras}$ (Vogel et al. 1988). Several lines of evidence have suggested that $p120^{GAP}$ may act either as an upstream regulator of $p21^{ras}$ or as a downstream effector molecule in the *ras*-mediated signal transduction pathway (for review see Downward 1992; Hall 1992). The addition of purified $p120^{GAP}$ or $p21^{ras}$ to isolated muscle cell membranes inhibited the G-protein-mediated coupling of K^+ channels to muscarinic cholinergic receptors as measured by patch clamping. When neutralizing anti-Ras antibodies were added, inhibition by $p120^{GAP}$ was lifted. Conversely, an antibody to GAP inhibited the effect of Ras, suggesting that Ras and GAP act together (Yatani et al 1990; Martin et al. 1992;). The role of $p120^{GAP}$ as a negative regulator of Ras function is suggested by its suppressing effects on oncogene-transformed cells (see below). There are a number of ways by which $p120^{GAP}$ can be regulated or modified. Regulatory mechanisms include phosphorylation on tyrosines by receptor and nonreceptor tyrosine kinases, complex formation with activated receptors such as PDGF receptor, interactions with two other phosphoproteins of 62 kDa and 190 kDa and certain lipids (Molloy et al. 1989, 1992; Ellis et al. 1990; Tsai et al. 1990; for review see Downward 1992).

The role of GAP in *ras*-mediated neoplastic transformation was investigated by cotransfection into NIH/3T3 cells of normal or mutant *ras* genes placed under the control of a murine retroviral LTR and a human GAP cDNA clone (Zhang et al. 1990). Transfection of a three to sixfold excess of GAP cDNA resulted in inhibition of focus formation (average inhibition

84%). In contrast, v-*ras* mediated transformation was not significantly affected by GAP cotransfection (average inhibition 13%). Morphological revertants were isolated following transfection of GAP cDNA into c-*ras* transformed cell lines. The level of exogenous GAP protein, rather than that of p21^{ras} correlated with the morphological reversion of *ras* transformed NIH/3T3 cells. The authors concluded that GAP is an upstream regulator of *ras,* exerting its effect by reducing the level of active GTP-bound Ras protein (Zhang et al. 1990). Similar effects of GAP were reported for cells transformed by genes which appear to require functional Ras for exerting their phenotypic effects. Overexpression of translation initiation factor eIF-4E was shown to result in tumorigenic transformation of rodent fibroblasts. p21^{ras} is involved in this process (Lazaris-Karatzas et al. 1992). Forced overexpression of GAP to a six to eight times elevated level compared to control eIF-4E transformed cells resulted in a flat morphology, loss of anchorage-independent growth and tumorigenicity in nude mice. At the same time, the levels of GTP-bound Ras were diminished to those found in immortalized nontumorigenic rat embryo fibroblasts. Two groups have reported that GAP inhibits neoplastic transformation by c-*src* or v-*src* (DeClue et al. 1991; Nori et al. 1991). *Src*-mediated transformation is dependent on endogenous Ras activity (Smith et al. 1986). Focus formation was reduced by about 80% when a plasmid encoding full-length GAP was cotransfected with the c-*src* oncogene. The suppressing activity of molecular clones encoding only the N-terminal amino acids 1–987 or C-terminal amino acids 685–1047 of GAP was analyzed in the same transformation system. Inhibition of c-*src* induced morphological transformation was even enhanced by the C-terminal GAP clone, while the N-terminal clone was inactive. Since the N-terminal GAP clone lacked part of the catalytic domain required for GTPase stimulation of p21^{ras}, it was concluded that the mechanism of inhibition is due to the catalytic downregulation of Ras by GAP. All revertants had unaltered levels of pp60^{src} expression or of protein-tyrosine phosphorylation (De Clue et al. 1991). GAP-overexpressing NIH/3T3 cells were infected with a v-*src* mutant virus that is temperature-sensitive for transformation. Afterwards, the dependence of transformation on the *src* oncogene of control cells and of GAP-overexpressing cells was assayed at the permissive (35.5°C) and the restrictive (39.5°C) temperatures. The control cells converted to morphological transformants when grown at the permissive temperature and remained normal at the restrictive temperature, while GAP-overexpressing cells retained their normal flat morphology at either 35.5 or 39.5°C (Nori et al. 1991).

Further evidence for the function of rasGAP as a negative regulator of Ras was obtained from experiments using expression vectors encoding plasma membrane-targeted rasGAP (Huang et al. 1993). To achieve association of

GAP with the plasma membrane, Ras C-terminal motifs sufficient for this localization were cloned onto the C-terminus of rasGAP. Coexpression of oncogenic *ras* and the membrane-targeted GAP in NIH/3T3 cells resulted in suppression of transformation. In addition, proliferation of untransformed NIH/3T3 cells was abrogated in the absence of the oncogene.

6.4 *NF*-1 (Neurofibromin)

Several lines of evidence have indicated that the genetic locus associated with von Recklinghausen neurofibromatosis type 1 (NF-1) harbors a tumor suppressor gene. The disease primarily affects tissues derived from the neural crest. Diagnostic features of NF-1 include pigmented skin patches (cafe au lait spots), skin neurofibromas containing multiple cell types such as Schwann's cells and axons, as well as Lisch nodules, a developmental abnormality of the iris. Less common features of NF-1 are malignant tumors of the central and peripheral nervous systems (for review see Ponder 1990). The NF-1 locus was assigned to the proximal long arm of human chromosome 17 by linkage analysis (Barker et al. 1987; Seizinger et al. 1987; Goldgar et al. 1989) and by physical mapping (Fountain et al. 1989; O'Connell et al. 1989; Yagle et al. 1990). The putative gene was localized to a DNA region comprising a few hundred kilobases. DNA samples of patients carrying translocations in the chromosomal region 17q11.2 were instrumental in the identification of the NF-1 gene that was reported independently by three groups. Viskochil et al. (1990) used a DNA sequence conserved in the human and rodent genomes that spanned a translocation breakpoint to identify a cDNA clone specific for a large 11-kb mRNA. In the same way, Cawthon et al. (1990) identified cDNA clones from the translocation breakoint region and a corresponding genomic sequence. Once the boundaries of introns and exons were identified, oligonucleotide primers were synthesized and used to compare amplified exons from normal donors and NF-1 patients by single-strand conformation polymorphism analysis. DNA fragments exhibiting an altered electrophoretic mobility were sequenced and searched for mutations. Nucleotide exchanges resulting in amino acid substitutions were identified in NF-1 patients, suggesting that the gene located at the translocation breakpoint was indeed the NF-1 gene. Using chromosome jumping techniques and recombinant yeast artificial chromosomes, Wallace et al. (1990) identified an approximately 13-kb mRNA transcript from the chromosomal region that is translocated in NF-1 patients. The gene was shown to be interrupted by the t(1;17) translocation. In addition, an insertion into the gene was identified in another patient resulting in an abberrant mRNA transcript. Cloning of the complete coding region of the NF-1 transcript revealed an open reading frame of 2818 amino acids. The gene spans approximately 300 kb and is organized into at least 30 exons (Marchuk et al. 1992). Several other genes were embedded in the region spanned by the novel gene and transcribed from the opposite DNA strand. Based on the mutational analysis, those genes were excluded as candidates for the NF-1 gene (Wallace et al. 1990; Xu et al. 1990; Cawthon et al. 1990, 1991).

6.4.1 Functional Relationship of NF-1 and Mammalian GAP

While the sequence of partial NF-1 cDNA clones initially did not reveal any similarities to known proteins, identification of additional 5' sequences permitted detection of a relevant protein domain. A sequence of 360 amino acid residues of the predicted polypeptide showed significant similarity to the C-terminal catalytic domain of human and bovine GAP proteins (Xu et al. 1990) and of the yeast IRA1 ("inhibitor of RAS") product (Buchberg et al. 1990). The overall sequence conservation of 23% between human NF-1 and yeast IRA1 is quite remarkable in that it extends over a region of more than 1500 amino acids. Later it was shown by both genetic and biochemical assays that the product of the *NF-1* gene interacts with p21^{ras}. A fragment of the NF-1 cDNA encoding the GAP-related domain (NF-1 GRD) was cloned in *Baculovirus* transfer or yeast expression vectors and expressed in Sf9 insect cells or in yeast, respectively. The protein was purified and assayed for its ability to stimulate human p21^{N-ras} GTPase activity. GTPase activity of normal *ras* was stimulated by NF-1 GRD, while two different activated *ras* mutant proteins were unaffected (Martin et al. 1990; Ballester et al. 1990; Xu et al. 1990). The affinity of NF-1 GRD for p21^{ras} was 20 times higher than that of rasGAP, suggesting that NF-1 is a potential regulator of p21^{ras} activity, especially when cellular p21^{ras} concentrations are low (Martin et al. 1990).

Further evidence for a role of *NF-1* as a regulator of *ras* genes was derived from complementation experiments in yeast. *IRA1* and *IRA2* genes of *S. cerevisiae* function upstream of *RAS1* and *RAS2* genes. Their products downregulate *RAS* activity by stimulating the GTPase activity of *RAS*. *ira*$^{-}$ mutants are characterized by sensitivity toward heat shock and lack of glycogen accumulation (Tanaka et al. 1989, 1990a,b, 1991). Similar phenotypes can be detected in yeast cells expressing mutated *RAS2* genes. Expression of NF-1 GRD cloned in a yeast expression vector was shown to suppress the heat shock sensitive phenotype of *ira*$^{-}$ mutants. Moreover, the ability to store glycogen was restored by NF-1 GRD (Xu et al. 1990; Martin et al. 1990; Ballester et al. 1990). Thus, NF-1 GRD functions similarly to mammalian GAP in the complementation of yeast mutant defects. The biochemical similarity of the rasGAP, NF-1 and IRA2 proteins is also reflected in their sensitivity toward inhibition of GTPase-activation by arachidonic acid (Golubic et al. 1991, 1992). The product of the *NF-1* gene does not contain SH2 or SH3 domains (Marchuk et al. 1992). Therefore, it is unlikely to be regulated through the same tyrosine kinases that are involved in the control of $p120^{GAP}$.

Antisera were raised to the bacterially expressed GAP-related domain (DeClue et al. 1991; Hattori et al. 1992), to the C-terminus (Hattori et al. 1992), and to the the regions of the *NF-1* gene product that flank the GRD (Daston et al. 1992). The size of the protein detected by these various antibodies was estimated between 220 and 320 kDa (DeClue et al. 1991, Hattori et al. 1992; Daston et al. 1992; Golubic et al. 1992; The et al. 1993). The protein was detectable in the particulate subcellular fraction in brain, but not in spleen, thymus, kidney, liver, lung, and heart (Hattori et al. 1992). Golubic et al. (1992) also described abundant expression in the brain; however, most of immunoprecipitated neurofibromin was present in the soluble rather than the particulate fraction. Immunostaining of tissue sections from the nervous system indicated that neurons, oligodendrocytes, and nonmyelinating Schwann's cells express the *NF-1* gene product (also named neurofibromin), while astrocytes and myelinating Schwann cells do not express detectable amounts (Daston et al. 1992).

Two alternatively spliced exons have been described for the human *NF-1* gene (Xu et al. 1990; Marchuk et al. 1991). When cDNA sequences of the NF-1 GRD were examined from several brain tumor cell lines, a novel form of NF-1 GRD cDNA was isolated. This cDNA contains an extra 63 bases encoding 21 amino acids inserted into the region of the GAP-related domain (Nishi et al. 1991; Suzuki et al. 1991). Type I transcripts without the insertion were found to be predominantly expressed in undifferentiated cells such as fetal brain, while type II transcripts containing the 63-bp insertion were detected in differentiated tissue and in most primary brain tumors analyzed (Teinturier et al. 1992). On treatment with the differentiation-inducer retinoic acid (RA), the predominant NF-1 GRD expression changed from type I to type II in human SH-SY5Y neuroblastoma cells. An astroytoma cell line that does not respond to RA showed no notable change in NF-1 GRD isoform expression (Nishi et al. 1991). It is not known how the catalytic properties of the protein are changed by the amino acid insertion; however, analysis of the predicted secondary structure indicated that the activity may be modified (Nishi et al. 1991; Suzuki et al. 1991). Although the predicted additional NF-1 peptide exhibits 42% amino acid identity with nucleoside triphosphatase I, its functional significance is not known (Teinturier et al. 1992). Recently it was shown that both forms of NF-1 GRD were able to complement loss of *IRA* function when expressed in yeast (Andersen et al. 1993). Another NF-1 cDNA encoding a truncated protein was described recently that might also be involved in the regulation of activity in cellular signal transduction. The predicted protein of 551 amino acids lacks the GAP/IRA-like region, but shares the NH_2-terminal 547 amino acid residues with authentic NF-1 protein (Suzuki et al. 1992).

$p21^{ras}$ can be regulated by rasGAP and NF-1 (neurofibromin) GAP in the same cell. Although a significant proportion of the NF-1 protein was found in monomeric configuration in the P100 membrane fraction of cells, it is associated with other cellular proteins to form complexes of 600–800 kDa as shown by glycerol gradient centrifugation after gentle cell lysis (DeClue et al. 1991). In contrast, the GAP protein was found mainly in the cytoplasm (Trahey and McCormick 1987) and may translocate to the plasma membrane upon phosphorylation on tyrosine residues (Ellis et al. 1990; Molloy et al. 1989). This difference between NF-1 and GAP proteins with respect to subcellular localization has important implications for the regulation of the target molecule $p21^{ras}$. Following tyrosine kinase receptor stimulation or transformation, Ras may be regulated in an inducible manner by the GAP protein able to shuttle between cytosol and plasma membrane. Alternatively, Ras may be controlled constitutively by constantly membrane-associated NF-1 protein (DeClue et al. 1991). Bollag and McCormick (1991) have compared GAP activities on normal, activated and effector mutants of $p21^{ras}$ by measuring the intrinsic GTPase activities, the binding affinities to ras-GAP and NF-1 GAP, and the stimulated GTPase activities. They found that both GAP proteins bind oncogenic *ras* proteins, but do not stimulate their GTPase activity. Moreover, NF-1 GAP binds to p21 *ras* up to 300 times more efficiently than rasGAP. In extracts of mammalian cells, both GAP activities are present. GAP activities can be selectively inhibited by certain lipids that play a role as second messengers in signal transduction. Furthermore, the detergent *n*-dodecyl-β-D-maltoside specifically inhibits NF-1 GAP. GAP activity in brain cell extracts was described that was not inhibited by anti-NF-1 and anti-GAP antibodies, suggesting the presence of yet another protein with GAP activity (Hattori et al. 1992). The neurofibromin-GRD peptide exhibited a 20-fold higher affinity for $p21^{Ha-ras}$ than GAP (Martin et al. 1990). Apart from the differences in the stimulation of the intrinsic GTPase activity of Ras, NF-1 GAP (neurofibromin) and rasGAP can be distinguished by the nature of their physical interaction with the proto-oncogene product (DiBattiste et al. 1993).

6.4.2 Mutations in Human Tumors and Abnormal Regulation of $p21^{ras}$

Malignant schwannoma cell lines derived from tumors removed from patients diagnosed with NF-1 expressed low to almost undetectable levels of NF-1 protein. The cell lines expressed normal levels of $p120^{GAP}$ and $p21^{ras}$.

In spite of a high GTPase activity indicative of normal *ras* gene function and of normal p120^{GAP} levels, these cell lines contain about 50% of Ras in the GTP-bound (i.e., activated) state which represents a tenfold elevation when compared to the level of Ras-GTP in normal cells (DeClue et al. 1992; Basu et al. 1992). A cDNA construct containing the catalytic domain of GAP linked to the *neo* gene was introduced by transfection into the ST88-14 schwannoma line that expressed very little NF-1 protein. Many of the transfected cell clones showed a flat, nonrefractile morphology. Some of the revertants tested grew poorly in semisolid agar medium and exhibited a reduced ability to form colonies. The level of GTP bound to Ras was reduced to that of normal control cells. Thus, the reversion of ST88-14 cells was most likely a direct consequence of the reduction in Ras-GTP due to expression of the GAP catalytic protein domain (DeClue et al. 1992). Since Ras confers a positive proliferative signal on these cells, NF-1 GRD appeared to be a major negative growth regulator in those neural crest-derived cells (Basu et al. 1992). These results demonstrated that an aberrant Ras signaling cascade is an important mediator of malignancy in neural crest derived tumors that are devoid of *ras* mutations (Bos 1989; for review see also Bollag and McCormick 1992).

Li et al. (1992) identified a somatically acquired point mutation affecting codon 1423 of the *NF-1* gene in one single sample from each of 22 colon adenocarcinomas, 10 anaplastic astrocytomas, and 28 peripheral blood samples from patients with myelodysplastic syndrome. The mutations altered a lysine residue in the catalytic domain of neurofibromin which is invariant among GAP-related proteins. A mutation in the same codon was identified in a family with neurofibromatosis type 1. To assay the functional consequences of these mutations, genetically engineered NF-1 cDNA encoding the GRD was expressed in a *Baculovirus* transfer vector. The mutant NF-1 proteins harboring either a Glu or a Gln residue instead of a Lys residue were purified from Sf9 insect cells and their GTPase-stimulating activity assayed. The activities of the Lys to Gln and Lys to Glu mutants were reduced by approximately 400- and 200-fold, respectively, while the affinities of either N-ras-GTP or Ha-ras-GTP for mutant and wild-type NF-1 GRD remained alike (Li et al. 1992). Legius et al. (1993) described the homozygous inactivation of the *NF-1* gene in a fibrosarcoma from a patient with neurofibromatosis type 1. Loss of heterozygosity was found for all chromosome 17 polymorphisms analyzed. The remaining chromosome 17 had a 200-kb deletion of the *NF-1* gene. Further evidence for NF-1 gene mutations in tumors not commonly found in neurofibromatosis type I patients was provided by analysis of neuroblastoma cell lines. Four out of ten cell lines expressed little or no neurofibromin, two of those showed abberrant restriction fragments within the *NF-1* gene (The et al. 1993). Interestingly, NF-1 deficient neuroblastoma cell lines showed high levels of Ras-GTP. Their proliferation could be inhibited by overexpression of p120^{GAP}. The authors speculated that NF-1 is required in a *ras*-mediated differentiation pathway which is blocked by mutations in tumor cells. A homozygous deletion of most of the NF-1 was detected in one out of eight melanoma cell lines analyzed. NF-1 cDNA and protein were not detectable in this cell line (Andersen et al. 1993).

Overexpression of NF-1 GRD cDNA encoding amino acid residues 1194–1531 in v-Ha-*ras* transformed NIH/3T3 fibroblasts significantly reduced their ability to proliferate without anchorage. With the help of a series of truncated NF-1 GRD constructs the region of the protein required for stimulation of c-Ha-*ras* GTPase was further narrowed down. A subdomain of 91 amino acids ranging from residue 1441–1531 activated the c-Ha-*ras* GTPase as shown by nucleoside release assays. It is not yet known whether these amino acids are sufficient to confer the anchorage-dependent phenotype on *ras* transformed cells (Nur-E-Kamal et al. 1993). As indicated by Downward (1992), it not yet formally proven that the full-length NF-1 protein has GAP activity towards p21^{ras}.

6.5 Ha-*ras*

The transfection of plasmid DNA containing both the nonmutated Ha-*ras* proto-oncogene and the mutant Ha-*ras* gene from T24 bladder carcinoma cells (Reddy et al. 1982) into immortalized rat 208F cells resulted only in a low frequency of transformed colonies. Most of the colonies were morphologically normal, anchorage-dependent, and nontumorigenic. The normal phenotype was induced even in stable mutant *ras* transfectants, into which the normal *ras* gene was introduced subsequently. Suppressed colonies expressed more normal than mutant p21^{ras} (Spandidos and Wilkie 1988). Introduction of the normal Ha-*ras* gene under the control of its own promoter had only little effects on the tumorigenic phenotype of the human bladder carcinoma cell line T24. When the endogenous Ha-*ras* promoter was replaced by a metallothioneine promoter, a marked, but unstable suppression of the transformed phenotype was observed. The authors suggest that the suppression is caused by competition of the normal Ha-*ras* gene product with the mutant protein for cellular targets, for example, the GAP protein. It is certainly interesting to note that cancer cell lines often express only mutant alleles of *ras* genes (Feinberg et al. 1983; Santos et al. 1984; Capon et al. 1983). Loss of the normal *ras* allele and retention of a mutated allele is also frequently found in primary human tumors (for review see Bos 1989). There are cell lines and tumors that coexpress a mutant and a normal *ras* allele (Shen et al. 1987; Paterson et al. 1987, Bos 1989). It is interesting to note that the replacement of one normal allele of the Ha-*ras* gene by a mutated copy did not result in neoplastic transformation of immortalized Rat1 cells provided that the mutant allele was not overexpressed (Finney and Bishop 1993). The results reported by Spandidos and Wilkie (1988) are in contrast to the finding that overexpression of the normal *ras* proto-oncogene results in neoplastic transformation of immortalized mouse NIH/3T3 cells (Chang et al. 1992). The reason for the differential sensitivity of cell lines toward introduction of the proto-oncogene is not understood. Possibly, sensitive cells such as NIH/3T3 have sustained alterations in critical regulatory circuits resulting in uncontrolled proliferation due to an enhanced Ras-GTP pool. Mutations in upstream or downstream effectors may stimulate mitogenic signaling in these cells. Preneoplastic NIH/3T3 cells and rat 208F cells differ in their ability to resist *ras* oncogene-mediated transformation. Cell fusion studies revealed that rat 208F cells were able to suppress the neoplastic phenotype when hybridized to Ha-*ras*-transformed rat FE-8 cells. In contrast, mouse NIH/3T3 cells had lost suppressive activity (Iten et al. 1989).

6.6 *Thy*-1

Sugimoto et al. (1991) have examined the levels of several cell surface proteins in oncogene-transformed NIH/3T3 fibroblasts using fluorescein isothiocyanate (FITC) conjugated antibodies and a fluorescence-activated cell sorter. The proteins included fibronectin, collagen type IV, laminin, Thy-1, and the histocompatibility antigen H2 K. The levels of Thy-1 protein and its mRNA were significantly reduced in DT cells and cell line 635, two derivatives of NIH/3T3 containing two or one copy of KiMSV, respectively. Thy-1 is a cell surface glycoprotein anchored to the cell membrane by glycophosphatidyl-inositol (Low et al. 1985; Tse et al. 1985). The function of Thy-1 is still unknown. It has been implicated to play a role in cell-cell recognition (Williams 1985), particularly in development (Morris 1985; Stern 1973), as well as in proliferation and differentiation of different cell types (Chen et al. 1987; Kollias et al. 1987). The level of Thy-1 mRNA in untransformed NIH/3T3 and *ras* transformed derivatives is probably regulated by a degradative pathway rather than by its synthesis as shown by nuclear runoff transcription assays. Since there was no absolute correlation between Thy-1 downregulation and expression of malignant properties in *ras* transformed cells and phenotypic revertants (Noda et al. 1983; 1989), Sugimoto et al. determined the effect of Thy-1 overexpression on the susceptibility of NIH/3T3 cells toward transformation by KiMSV. Following Ki-MSV-infection, Thy-1 negative NIH/3T3 cells produced 3-times as many colonies in semisolid agar medium as Thy-1-positive cells. Transfection of *Thy-1* cDNA into *ras*-transformed NIH/3T3 resulted in the partial inhibition of anchorage-independent growth. In addition, *Thy-1* overexpression in *ras* transformed A1 cells resulted in lower colony-forming efficiency in soft agar and reduced tumor formation in nude mice. In contrast, DT cells did not allow a high *Thy-1* expression after transfection of *Thy-1* cDNA and continued to express the transformed phenotype.

6.7 *gas*-1

The expression of a number of genes is increased upon serum starvation or density inhibition of normal fibroblasts. Several growth arrest-specific genes (*gas* genes) were isolated by differential cDNA hybridization (Schneider et al. 1988; Manfioletti et al. 1990). The product of the *gas*-1 gene is an integral membrane protein of 384 amino acids containing two putative transmembrane domains flanking an extracellular region encompassing amino acids 75–363 (Del Sal et al. 1992). When *gas*-1 was overexpressed from a constitutive promoter in quiescent cells, the serum-induced transition from the G_0 to the S phase of the cell cycle is blocked. However, the ectopic expression of *gas*-1 does not inhibit induction of immediate early proto-oncogenes *fos* and *jun*. Normal cycling cells were able to respond to an increased expression of *gas*-1 by exiting from the cell cycle. Ectopic expression of *gas*-1 driven by an SV40 promoter following microinjection of plasmid DNA resulted in a significant inhibition of DNA synthesis in NIH/3T3 cell lines transformed by either v-*ras*, v-*src*, v-*fos*, or v-*myc* oncogenes. SV40 transformed NIH/3T3 cells were insensitive toward *gas*-1 inhibition. Reduced expression of *gas*-1 was reported in Ki-*ras* transformed NIH/3T3 cells and primary mouse fibroblasts (Cairo et al. 1992). However, high levels of *gas*-1 mRNA were detected in rapidly proliferating fibrosarcoma and rhabdomyosarcoma maintained as xenografts on syngeneic mice. Therefore, the authors concluded that *gas*-1 expression was insufficient to inhibit proliferation and revert the tumorigenic phenotype of these cells.

The *gas*-2 gene encodes a protein of an apparent molecular mass of 36 kDa that is a component of the microfilament system. The *gas*-2 protein localizes with actin fibers, at the cell boundary and along stress fibers in growth-arrested NIH/3T3 fibroblasts. While *gas*-2 expression increased in serum-starved nontransformed cells, expression did not increase in cells transformed by v-*ras*, v-*src*, v-*fos*, and v-*myc* oncogenes (Brancolini et al. 1992). The effect of *gas*-2 overexpression in oncogene-transformed cells after introduction of the gene has not been reported.

6.8 *p53*

Molecular analysis of human cancers has recently provided convincing evidence for the multistep process of tumorigenesis that was initially suggested by Foulds (1958). Particularly, the genetic lesions occurring in colorectal tumors have been analyzed in much detail. As summarized by Fearon and Vogelstein (1990), the accumulation of at least four or five genetic alterations is thought to be required for the initiation of neoplasia and the gradual progression of normal colonic epithelial cells into a malignant tumor cell population. These alterations include the mutational activation of proto-oncogenes, especially of Ki-*ras*, and the functional inactivation of tumor suppressor genes such as *p53, MCC/APC,* and *DCC.* Similar combinations of genetic changes may occur in other types of human tumors as well. Many human tumors have been screened for mutations in the *p53* tumor suppressor gene and in *ras* oncogenes (for review see Levine et al. 1991; Hollstein et al. 1991; Bos 1989). To date, mutation of *p53* is regarded as the most frequently found genetic alteration in human tumors. The p53 phosphoprotein was originally detected as a tumor antigen of 393 amino acids in cells transformed by SV40. The protein was shown to form physical complexes with the large T antigen of SV40 and with the oncogene products of other DNA tumor viruses. The p53 protein can bind to double-stranded DNA as well as to RNA and can act as a regulator of transcription. It can stimulate the transcription of some genes such as the muscle creatine kinase gene, but also repress the transcription of other genes such as the c-*myc* oncogene. Mutations in the *p53* gene that are clustered into hot spots at least in some tumors result in an increase of the half-life of its protein product and accumulation in tumor cells. However, the normal wild-type protein may be accumulated also in normal cells after exposure to DNA damaging agents such as irradiation or drugs (for review see Lane 1992).

Mutations in the *p53* or one of the different *ras* genes can exist in the same cell, consequently they provide an interesting model to study the functional consequences of suppressor gene inactivation and oncogene activation. Mutant and wild-type (wt) *p53* genes exerted fundamentally different activities in cotransfection experiments with activated *ras* genes. Some, but not all mutated *p53* genes were able to act as immortalizing oncogenes in cotransformation experiments using rat embryo fibroblasts as recipients (Levine et al. 1991; Michalovitz et al. 1991; Farrell et al. 1991). Moreover, the introduction of mutant *p53* genes into tumorigenic cells that had lost both wild-type alleles resulted in an enhacement of malignant properties (Michalovitz et al. 1991; Zambetti et al. 1992; Gerwin et al. 1992). However, wild-type *p53* genes appeared to inhibit the expression of neoplastic transformation.

6.8.1 Negative Growth Regulation by p53
in the Presence of Activated ras Oncogenes

The p53 expression plasmids initially used for transfection assays contained mutations and thus did not resemble the true wild-type gene. Wild-type p53 expression plasmids were shown to inhibit the formation of transformed foci derived from normal rodent fibroblasts when introduced simultaneously with cooperating oncogenes such as *myc* plus *ras* and adenovirus E1A plus *ras* (Eliyahu et al. 1989; Finlay et al. 1989). Clones displaying a transformed morphology were infrequent in these transfections. Although all of those contained the *p53* gene integrated in their genome, these clones either failed to express the protein or expressed a mutated form of it (Finlay et al. 1989). In contrast to normal embryo fibroblasts, an established rat cell line, Rat-1, was only marginally sensitive toward inhibition by the normal *p53* gene. Cotransfection of *ras* or *src* oncogenes together with wt *p53* also did not result in a significant reduction in the number of transformants (Eliyahu et al. 1989).

The growth-inhibitory activity of wt *p53* was also investigated in human carcinoma cell lines. Baker et al. (1990) have introduced wt *p53* into SW 480 colon carcinoma cell line harboring an activated Ki-*ras* oncogene. Colony formation of wt *p53* transfectants was reduced five-to tenfold as compared to mutant *p53* transfectants. Colonies obtained after wt *p53* transfection contained rearranged or deleted exogenous *p53* DNA sequences and *p53* mRNA expression was not detectable. The carcinoma cells expressing the wt *p53* gene were unable to progress through the cell cycle. The growth inhibition of wt *p53* was not due to unspecific toxicity of its protein product, since the growth of epithelial cells derived from a benign colorectal tumor (VACO 235) was unaffected (Baker et al. 1990). The antiproliferative activity of a gene on chromosome 17 was confirmed by another group in an indirect way. Goyette et al. (1992) introduced chromosomes 5, 15, 17, or 18 into SW 480 colon carcinoma cells via microcell fusion. Microcell hybrids containing chromosome 18 produced slowly growing tumors in only some of the animals injected, whereas chromosome 5 hybrids were strongly suppressed for tumorigenicity and exhibited morphological alterations. Transfer of chromosome 17 into SW 480 cells did not yield viable colonies suggesting that this chromosome carries an active growth suppressor. Transfer of chromosome 15, which was regarded as a chromosome irrelevant for cell growth control, did not induce any phenotypic changes in SW 480 target cells. Sharma et al. (1993) compared the antiproliferative effect of wt *p53* on human EJ bladder carcinoma, SW 480 colon carcinoma, and HT 1080 fibrosarcoma cell lines harboring activated Ha-*ras*, Ki-*ras*, and N-*ras* oncogenes, respectively. As shown by both transient and stable transfections, wt *p53* was antiproliferative irrespective of the status of the endogenous *p53* gene. These studies supported the notion that wt *p53* exerts a strong antiproliferative effect on tumor cells expressing Ki-*ras*.

The simultaneous effect of wt *p53* and oncogenes on the proliferation of murine NIH/3T3 cells was investigated by Chen and Defendi (1992). Expression of exogenous wt *p53* resulted in a more than 95% reduction of colony formation. This antiproliferative effect was partially abolished by cotransfection of activated human Ha-*ras*, since the yield of colonies inreased about tenfold. The wild-type *p53* gene had no effect on in vitro transformation in *ras*-transformed NIH/3T3 cells. Transfected cells continued to grow in low serum medium, to exhibit a transformed morphology and anchorage independence. In contrast, HPV18 E6 and c-*myc* oncogenes were able to overcome the antiproliferative but not the antitransforming effects of wt *p53*. The functional interaction of *ras* oncogenes and endogenous wt *p53* genes was investigated in reconstituted mouse prostate organs (Lu et al. 1992). Epithelial and mesenchymal cells were isolated from fetal mouse urogenital sinus and infected with recombinant retroviruses carrying marker genes and also *ras* or *myc* oncogenes. The manipulated cells were grafted into the ventral kidney capsule of mice, where they differentiated into prostatelike tissue (Thompson et al. 1989). In reconstituted mouse prostate, the *ras* oncogene caused hyperplastic growth (R cells), whereas the combinations of *ras* and *myc* genes (RM cells) resulted in the formation of carcinomas. Mutations in the endogenous *p53* gene were not found in these tumors. In RM cells, the expression of endogenous wt p53 was high, but very heterogenous. Cell cycle control and mitosis were perturbed, as indicated by the presence of micronuclei, giant and multinucleated cells. RM cells were completely unresponsive toward the growth-inhibitory activity of exogenous wt p53. Instead, R cells harbored *p53* mutations and were growth-arrested by exogenous wt *p53*. Thus, the susceptibility of *ras*-expressing cells toward wt *p53* is governed by the state of endogenous *p53*. The defective growth control function was restored by introduction of wt *p53* into R cells. The coexpression of *myc* bypassed the need for *p53* mutations in RM cells. Introduction of wt *p53* no longer had an antiproliferative effect.

7 Discussion and Perspectives

The reversion of *ras*-mediated transformation can be achieved by different means such as treatment with certain antibiotics and cytokines, expression of antisense RNA or oligonucleotides to the oncogene, or the forced expression of transformation suppressor genes. Most importantly, *ras* oncogene-expressing cells may loose their oncogenic properties in vivo and in vitro even if the oncoprotein level is unaltered. The expression of the neoplastic phenotype may be blocked by inactivation of an effector molecule (or molecules) essential for the transduction of the mitogenic signal such as in effector⁻ mutants or by the activation of an endogenous suppressor gene acting in a dominant manner. Cell lines that are resistant toward *ras* transforming activity may have sustained similar alterations as revertants obtained from transformed cells. Several genes have been identified that are directly or indirectly associated with the suppression of neoplastic transformation. From the study of those putative or proven growth suppressor genes, the following mechanisms of suppression have emerged: the expression of the *ras*-induced transformed phenotype may be perturbed at the level of the signal transduction cascade downstream of Ras, at the level of transcriptional control, and possibly at the level of cellular interactions with the extracellular matrix and with neighboring cells. Two genes, *Krev*-1 and *rsp*-1, were identified by expression cloning, the method that provides the most direct approach to molecular clones of transformation suppressor genes. However, both genes exerted their suppressing activity only when overexpressed in *ras*-transformed cells. This raises the question, whether expression of these genes at physiological levels would also confer any constraints on cellular growth. An important hint for a role as regulators of normal proliferation or differentiation would be the detection of inactivating mutations in tumorigenic cells. Mutations in the *rsp*-1 gene have not been reported and the search for alterations of *Krev*-1 in experimental and human tumors has been rather discouraging. In contrast, the *NF-1* gene encoding neurofibromin was isolated by positional cloning based on its association with von Recklinghausen neurofibromatosis type 1. Mutations or deletions were found in NF-1-associated tumors and in a few cell lines derived from other tumors. It is intriguing that neural crest derived tumors found in neurofibromatosis type 1 patients have not sustained oncogenic mutations in *ras* genes. However, loss of neurofibromin expression abolishes GAP activity and causes an elevation of activated Ras-GTP. Thus, a continuous mitogenic signal can be generated, although the GTPase activity of Ras itself is not affected by mutations. Similarly, pertubation of rasGAP activity could result in Ras activation as suggested by the inhibition of *ras*- or *src*-induced transfomation by overexpression of *ras* GAP. Unlike the *p53* tumor suppressor gene, the genes encoding rasGAP or neurofibromin (NF-1 GAP) represent a large target for mutations, due to their considerable size. For the same reason, both genes were unlikely to be detected by gene transfer and expression cloning. The important function of these proteins in Ras regulation warrants further screenings for mutations in tumors.

Several genes were specifically expressed in phenotypic revertants derived from *ras*-transformed cells. Reversible downregulation of gene expression during the conversion from the normal to the transformed state and reexpression upon reversion, or selective expression in revertants does not define a causal role in those phenotypic changes. Genes encoding different types of collagens or other components of the extracellular matrix were preferentially detected by differential hybridization techniques, since the corresponding transcripts are quite abundant in revertant cells and may represent the "tip of the iceberg." It is very unlikely that these genes were associated with the revertant state by coincidence, since they were repeatedly identified in independent sets of revertants. These genes, coined class II suppressor genes, may serve as important markers for the normal and the revertant state (Lee et al. 1991). It is interesting to note that the genes encoding collagens and lysyl oxidase (*rrg*) appear to be coordinately downregulated on *ras* transformation. Reexpression of some structural components of the extracellular matrix and of the enzyme lysyl oxidase mediating collagen and elastin cross-linking may have, at least indirect, growth-constraining effects on cells. Cellular proliferation and proto-oncogene activity can be modulated by cell shape (Folkman and Moscona 1978; Tucker et al. 1981; Kulesh and Greene 1986; Farrell and Greene 1992). This notion is supported by the finding that overexpression of the fibronectin receptor (Giancotti et al. 1990) and of vinculin (Fernandez et al. 1992) in spontaneously transformed cells restores normal growth control.

The *p53* gene serves as a paradigm for class I suppressor genes (Lee et al. 1991), since its expression in tumorigenic cells had profound inhibitory effects on cellular proliferation and on transformed phenotypes. P53 is an important transcriptional activator which probably regulates a number of genes associated with cell survival, proliferation and differentiation. Due to its predominantly nuclear localization, the p53 protein does not directly interfere with elements of the Ras signaling pathway. It may, however, interfere with several downstream targets of Ras. Thus, *p53* expression may result in growth arrest even in the presence of Ras, particularly as long as the level of expression is very high. The existence of different signaling pathways leading to growth stimulation would explain why some *ras*-expressing cells failed to respond to growth inhibition by p53. Alternatively, these cells may have sustained mutations in the p53 pathway.

The multistep process of malignant transformation is tightly associated with the inactivation of suppressor gene activity. On the basis of chromosome transfer studies, it was proposed that restoration of a single missing suppressor activity is sufficient to cause phenotypic reversion, even if several other lesions including *ras* mutations persist (Goyette et al. 1992). If this concept is validated by direct analysis of the genes involved, successful therapeutic intervention would require restoration of single defects only. The restoration of a mutated or deleted gene function requires reintroduction of the wild-type gene into cancer cells in vivo, an approach which might be feasible in the not too distant future. An alternative view is that the restoration of normal growth control requires the coordinate action of several genes. Since the reverted phenotype is inducible by certain drugs and antibiotics, it is tempting to speculate that the genetic program underlying the expression of the normal state can be reactivated pharmacologically.

Acknowledgements. The author's work is supported by Schweizerischer Nationalfonds, Schweizerische Krebsliga, and Krebsliga des Kantons Zürich. I am grateful to the members of my laboratory and Lewis Rowell for helpful discussions.

References

Aaltonen LA, Peltomäki P, Leach FS, Sistonen P, Pylkkänen L, Mecklin J-P, Järvinen H, Powell SM, Jen J, Hamilton SR, Petersen GM, Kinzler KW, Vogelstein B, De la Chapelle A (1993) Clues to the pathogenesis of familial colorectal cancer. Science 260:812–816

Andersen LB, Ballester R, Marchuk DA, Chang E, Gutmann DH, Saulino AM, Camonis J, Wigler M, Collins FS (1993a) A conserved alternative splice in the Vonrecklinghausen neurofibromatosis (NF1) gene produces 2 neurofibromin isoforms, both of which have GTPase-activating protein activity. Mol Cell Biol 13:487–495

Andersen LB, Fountain JW, Gutmann DH, Tarlé SA, Glover TW, Dracopoli NC, Housman DE, Collins FS (1993b) Mutations in the neurofibromatosis 1 gene in sporadic malignant melanoma cell lines. Nature Genet 3:118–121

Argiolas A, Pisano JJ (1983) Facilitation of phospholipase A2 activity by mastoparans, a new class of mast cell degranulating peptides from wasp venom. J Biol Chem 258:13697–13702

Baker SJ, Fearon ER, Nigro JM, Hamilton SR, Preisinger AC, Jessup JM, Van Tuinen P, Ledbetter DH, Barker DF, Nakamura Y, White R, Vogelstein B (1989) Chromosome 17 deletions and p53 gene mutations in colorectal carcinomas. Science 244:217–221

Baker SJ, Markowitz S, Fearon ER, Willson JKV, Vogelstein B (1990) Suppression of human colorectal carcinoma cell growth by wild-type p53. Science 249:912–915

Ballester R, Marchuk D, Boguski M, Saulino A, Letcher R, Wigler M, Collins F (1990) The NF1 locus encodes a protein functionally related to mammalian GAP and yeast IRA proteins. Cell 63:851–859

Balmain A, Ramsden M, Bowden GT, Smith J (1984) Activation of the mouse cellular Harvey-ras gene in chemically induced benign skin papillomas. Nature 307:658–660

Bar-Sagi D, Feramisco JR (1985) Microinjection of the ras oncogene protein into PC12 cells induces morphological differentiation. Cell 42:841–848

Barbacid M (1987) ras genes. Annu Rev Biochem 56:779–827

Barker D, Wright E, Nguyen K, Cannon L, Fain P, Goldgar D, Bishop DT, Carey J, Baty B, Kivlin J, Willard H, Waye JS, Greig G, Leinwand L, Nakamura Y, O'Connell P, Leppert M, Lalouel J-M, White R, Skolnick M (1987) Gene for von Recklinghausen neurofibromatosis is in the pericentric region of chromosome 17. Science 236:1100–1102

Baron HM, Bobrisheva IV, Varshaver NB (1992) The activated human c-Ha-ras-1 oncogene as a mutagen. Cancer Genet Cytogenet 62:15–20

Bassin RH, Noda M (1987) Oncogene inhibition by cellular genes. Adv Viral Oncol 6:103–127

Basu TN, Gutmann DH, Fletcher JA, Glover TW, Collins FS, Downward J (1992) Aberrant regulation of ras proteins in malignant tumour cells from type-1 neurofibromatosis patients. Nature 356:713–715

Benade LE, Talbot N, Tagliaferri P, Hardy C, Card J, Noda M, Najam N, Bassin RH (1986) Ouabain sensitivity is linked to ras-transformation in human HOS cells. Biochem Biophys Res Comm 136:807–814

Benito M, Porras A, Nebreda AR, Santos E (1991) Differentiation of 3T3-L1 fibroblasts to adipocytes induced by transfection of ras oncogenes. Science 253:565–568

Birchmeier C, Broek D, Wigler M (1985) RAS proteins can induce meiosis in Xenopus oocytes. Cell 43:615–621

Bokoch GM, Parkos CA, Mumby SM (1988) Purification and characterization of the 22,000-dalton GTP-binding protein substrate for ADP-ribosylation by botulinum toxin G22 K. J Biol Chem 263:16744–16749

Bokoch GM, Quilliam LA, Bohl BP, Jesaitis AJ, Quinn MT (1991) Inhibition of Rap1A binding to cytochrome-b558 of NADPH oxidase by phosphorylation of Rap1A. Science 254:1794–1796

Bokoch GM (1993) Biology of the Rap proteins, members of the ras superfamily of GTP-binding proteins. Biochem J 289:17–24

Bollag G, McCormick F (1991) Differential regulation of rasGAP and neurofibromatosis gene product activities. Nature 351:576–579

Bollag G, McCormick F (1992) RAS regulation-NF is enough of GAP. Nature 356:663–664

Bos JL, Verlaan-de Vries M, Jansen AM, Veeneman GH, van Boom JH, Van der Eb AJ (1984) Three different mutations in codon 61 of the human N-ras gene detected by synthetic oligonucleotide hybridization. Nucleic Acids Res 12:9155–9163

Bos JL (1989) Ras oncogenes in human cancer: a review. Cancer Res 49:4682–4689

Bourne HR, Sanders DA, McCormick F (1991) The GTPase superfamily: Conserved structure and molecular mechanism. Nature 349:117–127

Brancolini C, Bottega S, Schneider C (1992) Gas2, a growth arrest-specific protein, is a component of the microfilament network system. J Cell Biochem 117:1251–1261

Buchberg AM, Cleveland LS, Jenkins NA, Copeland NG (1990) Sequence homology shared by neurofibromatosis type-1 gene and IRA-1 and IRA-2 negative regulators of the RAS cyclic AMP pathway. Nature 347:291–294

Buss JE, Quilliam LA, Kato K, Casey PJ, Solski PA, Wong G, Clark R, McCormick F, Bokoch GM, Der CJ (1991) The COOH-terminal domain of the Rap1A (Krev-1) protein is isoprenylated and supports transformation by an H-Ras:Rap1A chimeric protein. Mol Cell Biol 11:1523–1530

Bèranger F, Goud B, Tavitian A, de Gunzburg J (1991) Association of the Ras-antagonistic Rap1/Krev-1 proteins with the Golgi complex. Proc Natl Acad Sci USA 88:1606–1610

Cai H, Szeberenyi J, Cooper GM (1990) Effect of a dominant inhibitory Ha-ras mutation on mitogenic signal transduction in NIH 3T3 cells. Mol Cell Biol 10:5314–5323

Cairo G, Ferrero M, Biondi G, Colombo MP (1992) Expression of a growth arrest specific gene (gas-1) in transformed cells. Br J Cancer 66:27–31

Campa MJ, Chang K-J, yVedia LM, Reep BR, Lapentina EG (1991) Inhibition of ras-induced germinal vesicle breakdown in Xenopus oocytes by rap-1B. Biochem Biophys Res Comm 174:1–5

Cantley LC, Auger KR, Carpenter C, Duckworth B, Graziani A, Kapeller R, Soltoff S (1991) Oncogenes and signal transduction. Cell 64:281–302

Capon DJ, Chen EY, Levinson AD, Seeburg PH, Goeddel DV (1983a) Complete nucleotide sequences of T24 human bladder carcinoma oncogene and its normal homologue. Nature 302:33–37

Capon DJ, Seeburg PH, McGrath JP, Hayflick JS, Edman U, Levinson AD, Goeddel DV (1983b) Activation of Ki-ras2 gene in human colon and lung carcinomas by two different point mutations. Nature 304:507–513

Cardiff RD, Sinn E, Muller W, Leder P (1991) Transgenic oncogene mice. Tumor phenotype predicts genotype. Am J Pathol 139:495–501

Cawthon RM, Weiss R, Xu G, Viskochil D, Culver M, Stevens J, Robertson M, Dunn D, Gesteland R, O'Connell P, White R (1990a) A major segment of the neurofibromatosis type 1 gene: cDNA sequence, genomic structure, and point mutations. Cell 62:193–201

Cawthon RM, O'Connell P, Buchberg AM, Viskochil D, Weiss RB, Culver M, Stevens J, Jenkins NA, Copeland NG, White R (1990b) Identification and characterization of transcripts from the neurofibromatosis 1 region: The sequence and genomic structure of EVI2 and mapping of other transcripts. Genomics 7:555–565

Cawthon RM, Andersen LB, Buchberg AM, Xu G, O'Connell P, Viskochil D, Weiss RB, Wallace MR, Marchuk DA, Culver M, Stevens J, Jenkins NA, Copeland NG, Collins FS, White R (1991) cDNA sequence and genomic structure of EVI2B, a gene lying within an intron of the neurofibromatosis type 1 gene. Genomics 9:446–460

Chang EH, Furth ME, Scolnick EM, Lowy DR (1982) Tumorigenic transformation of mammalian cells induced by a normal human gene homologous to the oncogene of Harvey murine sarcoma virus. Nature 297:479–483

Chang EH, Miller PS, Cushman C, Devadas K, Pirollo KF, Ts'o POP, Yu ZP (1991) Antisense inhibition of ras p21 expression that is sensitive to a point mutation. Biochemistry 30:8283–8296

Chen C, Okayama H (1987) High-efficiency transformation of mammalian cells by plasmid DNA. Mol Cell Biol 7:2745–2752

Chen S, Botteri F, van der Putten H, Landel CP, Evans GA (1987) A lymphoproliferative abnormality associated with inappropriate expression of the Thy-1 antigen in transgenic mice. Cell 51:7–19

Chen TM, Defendi V (1992) Functional interaction of p53 with HPV18 E6, c-myc and H- ras in 3T3 Cells. Oncogene 7:1541–1547

Chiao PJ, Kannan P, Yim SO, Krizman DB, Wu T-A, Gallick GE, Tainsky MA (1991) Susceptibility to ras oncogene transformation is coregulated with signal transduction through growth factor receptors. Oncogene 6:713–720

Cho HY, Cutchins EC, Rhim JS, Huebner RJ (1976) Revertants of human cells transformed by murine sarcoma virus. Science 194:951–953

Ciardiello F, Sanfilippo B, Yanagihara K, Kim N, Tortora G, Bassin RH, Kidwell WR, Salomon DS (1988) Differential growth sensitivity to 4-cis-hydroxy-L-proline of transformed rodent cell lines. Cancer Res 48:2483–2491

Colicelli J, Field J, Ballester R, Chester N, Young D, Wigler M (1990) Mutational mapping of RAS-responsive domains of the Saccharomyces cerevisiae adenylyl cyclase. Mol Cell Biol 10:2539–2543

Compere SJ, Baldacci P, Sharpe AH, Thompson T, Land H, Jaenisch R (1989) The ras and myc oncogenes cooperate in tumor induction in many tissues when introduced into midgestation mouse embryos by retroviral vectors. Proc Natl Acad Sci USA 86:2224–2228

Contente S, Kenyon K, Rimoldi D, Friedman RM (1990) Expression of gene rrg is associated with reversion of NIH 3T3 transformed by LTR-c-Ha-ras. Science 249:796–798

Cooper GM, Okenquist S, Silverman L (1980) Transforming activity of DNA of chemically transformed and normal cells. Nature 284:418–421

Cox AD, Hisaka MM, Buss JE, Der CJ (1992) Specific isoprenoid modification is required for function of normal, but not oncogenic, ras protein. Mol Cell Biol 12:2606–2615

Craig RW, Sager R (1985) Suppression of tumorigenicity in hybrids of normal and oncogene-transformed CHEF cells. Proc Natl Acad Sci USA 82:2062–2066

Culine S, Olofsson B, Gosselin S, Honore N, Tavitian A (1989) Expression of the ras-related rap genes in human tumors. Int J Cancer 44:990–994

Cutler ML, Bassin RH, Zanoni L, Talbot N (1992) Isolation of rsp-1, a novel cDNA capable of suppressing v- Ras transformation. Mol Cell Biol 12:3750–3756

Daston MM, Scrable H, Nordlund M, Sturbaum AK, Nissen LM, Ratner N (1992) The protein product of the neurofibromatosis type 1 gene is expressed at highest abundance in neurons, Schwann cells, and oligodendrocytes. Neuron 8:415–428

DeClue JE, Cohen BD, Lowy DR (1991) Identification and characterization of the neurofibromatosis type 1 protein product. Proc Natl Acad Sci USA 88:9914–9918

DeClue JE, Papageorge AG, Fletcher JA, Diehl SR, Ratner N, Vass WC, Lowy DR (1992) Abnormal regulation of mammalian p21(ras) contributes to malignant tumor growth in Vonrecklinghausen (type-1) neurofibromatosis. Cell 69:265–273

Declue JE, Zhang K, Redford P, Vass WC, Lowy DR (1991) Suppression of src transformation by overexpression of full-length GTPase-activating protein (GAP) or of the GAP terminus. Mol Cell Biol 11:2819–2825

Delsal G, Ruaro ME, Philipson L, Schneider C (1992) The growth arrest-specific gene, gas1, is involved in growth suppression. Cell 70:595–607

Diamantis ID, Nair APK, Hirsch HH, Moroni C (1989) Tumor suppression involves down-regulation of interleukin 3 expression in hybrids between autocrine mastocytoma and interleukin 3-dependent parental mast cells. Proc Natl Acad Sci USA 86:9299–9302

Dibattiste D, Golubic M, Stacey D, Wolfman A (1993) Differences in the interaction of p21(c-Ha-ras)-GMP-PNP with full-length neurofibromin and GTPase-activating protein. Oncogene 8:637–643

Downward J, Graves JD, Warne PH, Rayter S, Cantrell DA (1990) Stimulation of p21ras upon T-cell activation. Nature 346:719–723

Downward J (1992) Regulatory mechanisms for ras proteins. Bioessays 14:177–184

Duronio V, Welham MJ, Abraham S, Dryden P, Schrader JW (1992) P21ras activation via hemopoietin receptors and c-kit requires tyrosine kinase activity but not tyrosine phosphorylation of p21ras GTPase activating protein. Proc Natl Acad Sci USA 89:1587–1591

Eliyahu D, Michalovitz D, Eliyahu S, Pinhasi-Kimhi O, Oren M (1989) Wild-type p53 can inhibit oncogene-mediated focus formation. Proc Natl Acad Sci USA 86:8763–8767

Ellis C, Moran M, McCormick F, Pawson T (1990) Phosphorylation of GAP and GAP-associated proteins by transforming and mitogenic tyrosine kinases. Nature 343:377–381

Farrell FX, Ohmstede C-A, Reep BR, Lapentina EG (1990) cDNA sequence of a new ras-related gene (rap2b) isolated from human platelets with sequence homology to rap2. Nucleic Acids Res 18:4281

Farrell PJ, Allan GJ, Shanahan F, Vousden KH, Crook T (1991) p53 is frequently mutated in Burkitt's lymphoma cell lines. EMBO J 10:2879–2887

Farrell RE, Greene JJ (1992) Regulation of c-myc and c-Ha-ras oncogene expression by cell shape. J Cell Physiol 153:429–435

Fearon ER, Vogelstein B (1990) A genetic model for colorectal tumorigenesis. Cell 61:759–767

Fearon ER, Cho KR, Nigro JM, Kern SE, Simons JW, Ruppert JM, Hamilton SR, Preisinger AC, Thomas G, Kinzler KW, Vogelstein B (1990) Identification of a chromosome 18q gene that is altered in colorectal cancers. Science 247:49–56

Feinberg AP, Vogelstein B, Droller MJ, Baylin SB, Nelkin BD (1983) Mutation affecting the 12th amino acid of the c-Ha-ras oncogene product occurs infrequently in human cancer. Science 220:1175–1177

Fernandez JLR, Geiger B, Salomon D, Sabanay I, Zoller M, Benzeev A (1992) Suppression of tumorigenicity in transformed cells after transfection with vinculin cDNA. J Cell Biol 119:427–438

Field J, Xu H-P, Michaeli T, Ballester R, Sass P, Wigler M, Colicelli J (1990) Mutations of the adenylate cyclase gene that block RAS function in Saccharomyces cervisiae. Science 247:464–467

Finlay CA, Hinds PW, Levine AJ (1989) The p53 proto-oncogene can act as a suppressor of transformation. Cell 57:1083–1093

Finney RE, Bishop JM (1993) Predisposition to neoplastic transformation caused by gene replacement of H-ras1. Science 260:1524–1527

Fischer TH, White II GC (1987) Partial purification and characterization of thrombolamban, a 22,000 dalton cAMP-dependent protein kinase substrate in platelets. Biochem Biophys Res Comm 149:700–706

Folkman J, Moscona A (1978) Role of cell shape in growth control. Nature 273:345349

Fountain JW, Wallace MR, Bruce MA, Seizinger BR, Menon AG, Gusella JF, Michels VV, Schmidt MA, Dewald GW, Collins FS (1989) Physical mapping of a translocation breakpoint in neurofibromatosis. Science 244:1085–1087

Franza BRjr, Maruyama K, Garrels JI, Ruley HE (1986) In vitro establishment is not a sufficient prerequisite for transformation by activated ras oncogenes. Cell 44:409–418

Frech M, John J, Pizon V, Chardin P, Tavitian A, Clark R, McCormick F, Wittinghofer A (1990) Inhibition of GTPase activating protein stimulation of Ras-p21 GTPase by the Krev-1 gene product. Science 249:169–171

Fry DG, Milam LD, Dillberger JE, Maher VM, McCormick JJ (1990) Malignant transformation of an infinite life span human fibroblast cell strain by transfection with v-Ki-ras. Oncogene 5:1415–1418

Fujita H, Suzuki H, Kuzumaki N, Müllauer L, Ogiso Y, Oda A, Ebisawa K, Sakurai T, Nonomura Y, Kijimoto-Ochiai S (1990) A specific protein, p92, detected in flat revertants derived from NIH/3T3 transformed by human activated c-Ha-ras oncogene. Exp Cell Res 186:115–121

Geiser AG, Anderson MJ, Stanbridge EJ (1989) Suppression of tumorigenicity in human cell hybrids derived from cell lines expressing different activated ras oncogenes. Cancer Res 49:1572–1577

Geiser AW, Der CJ, Marshall CJ, Stanbridge EJ (1986) Suppression of tumorigenicity with continued expression of the c-Ha-ras oncogene in EJ bladder carcinoma-human fibroblast hybrid cells. Proc Natl Acad Sci USA 83:5209–5213

Gerfaux J, Sergiescu D, Hamelin R, Joret A-M, Lallemand C (1990) A common cellular pathway for v-mos and v-Ki-ras is not required for v-Ki-ras-induced tumorigenicity in a nonmalignant, v-mos-expressing revertant cell. Mol Carcinog 3:103–113

Gerwin BI, Spillare E, Forrester K, Lehman TA, Kispert J, Welsh JA, Pfeifer AMA, Lechner JF, Baker SJ, Vogelstein B, Harris CC (1992) Mutant-p53 can induce tumorigenic conversion of human bronchial epithelial cells and reduce their responsiveness to a negative growth factor, transforming growth factor- beta(1). Proc Natl Acad Sci USA 89:2759–2763

Giancotti FG, Ruoslahti E (1990) Elevated levels of the alpha5/beta1 fibronection receptor suppress the transformed phenotype of Chinese hamster ovary cells. Cell 60:849–859

Gibbs JB, Marshall MS, Scolnick EM, Dixon RAF, Vogel US (1990) Modulation of guanine nucleotides bound to Ras in NIH 3T3 cells by oncogenes, growth factors, and the GTPase activating protein. J Biol Chem 265:20437–20442

Goldgar DE, Green P, Parry DM, Mulvihill JJ (1989) Multipoint linkage analysis in neurofibromatosis type I: An international collaboration. Am J Hum Genet 44:6–12

Golubic M, Tanaka K, Dobrowolski S, Wood D, Tsai MH, Marshall M, Tamanoi F, Stacey DW (1991) The GTPase stimulatory activities of the neurofibromatosis type 1 and the yeast IRA2 proteins are inhibited by arachidonic acid. EMBO J 10:2897–2903

Golubic M, Roudebush M, Dobrowolski S, Wolfman A, Stacey DW (1992) Catalytic properties, tissue and intracellular distribution of neurofibromin. Oncogene 7:2151–2159

Goyette MC, Cho K, Fasching CL, Levy DB, Kinzler KW, Paraskeva C, Vogelstein B, Stanbridge EJ (1992) Progression of colorectal cancer is associated with multiple tumor suppressor gene defects but inhibition of tumorigenicity is accomplished by correction of any single defect via chromosome transfer. Mol Cell Biol 12:1387–1395

Grand RJA, Owen D (1991) The biochemistry of ras p21. Biochem J 279:609–631

Graves JD, Downward J, Rayter S, Warne P, Turr AL, Glennie M, Cantrell DA (1991) CD2 antigen mediated activation of the guanine nucleotide binding proteins p21ras in human T lymphocytes. J Immunol 146:3709–3712

Graves JD, Downward J, Izquierdo M, Rayter S, Warne PH, Cantrell DA (1992) The growth factor IL-2 activates p21ras proteins in normal human T lymphocytes. J Immunol 148:2417–2422

Greenberger JS, Aaaronson SA (1974) Morphologic revertants of murine sarcoma virus transformed nonproducer BALB/3T3: Selective techniques for isolation and biologic properties in vitro and in vivo. Virology 57:339–346

Griegel S, Traub O, Willecke K, Schäfer R (1986) Suppression and re-expression of transformed phenotype in hybrids of Ha-ras1 transformed Rat-1 cells and early passage rat embryo fibroblasts. Int J Cancer 38:697–705

Guerrero I, Villasante A, Corces V, Pellicer A (1984a) Activation of a c-K-ras oncogene by somatic mutation in mouse lymphomas induced by gamma irradiation. Science 225:1159–1161

Guerrero I, Villasante A, D'Eustachio P, Pellicer A (1984b) Isolation, characterization, anmd chromosome assignment of mouse N-ras gene from carcinogen-induced thymic lymphoma. Science 225:1041–1044

Hafen E, Basler K, Edstroem J-E, Rubin GM (1987) Sevenless, a cell-specific homeotic gene of Drosophila, encodes a putative transmembrane receptor with a tyrosine kinase domain. Science 236 :55–63

Hajnal A, Klemenz R, Schäfer R (1993a) Suppression of ras-mediated transformation. Differential expression of genes encoding extracellular matrix proteins in normal, transformed and revertant cells. Adv Enzyme Regul 33:267–280

Hajnal A, Klemenz R, Schäfer R (1993b) Upregulation of lysyl oxidase in spontaneous revertants of H-ras transformed rat fibroblasts.

Hajnal A, Klemenz R, Schäfer R (1994) Subtraction cloning of H-Rev107, a gene specifically expressed in H-ras resistant fibroblasts. Cancer Res 53: 4670–4675

Hall A, Marshall CJ, Spurr NK, Weiss RA (1983) Identification of transforming gene in two human sarcoma cell lines as a new member of the ras gene family located on chromosome 1. Nature 303:396–400

Hall A (1990) ras and GAP – Who's controlling whom? Cell 61:921–923

Hall A (1992) Signal transduction through small GTPases – A tale of 2 GAPs. Cell 69:389–391

Hanahan D (1988) Dissecting multistep tumorigenesis in transgenic mice. Annu Rev Genet 22:479–519

Hariharan IK, Carthew RW, Rubin GM (1991) The Drosophila roughened mutation: activation of a rap homolog disrupts eye development and interferes with cell determination. Cell 67:717–722

Hart PA, Marshall CJ (1990) Amino acid 61 is a determinant of sensitivity of rap proteins to ras GTPase activating protein. Oncogene 5:1099–1101

Hata Y, Kikuchi A, Sasaki T, Schaber MD, Gibbs JB, Takai Y (1990) Inhibition of the ras p21 GTPase-activating protein-stimulated GTPase activity of c-Ha-ras p21 by smg p21 having the same effector domain as ras p21s. J Biol Chem 265:7104–7107

Hattori S, Maekawa M, Nakamura S (1992) Identification of neurofibromatosis type-I gene product as an insoluble GTPase-activating protein toward ras p21. Oncogene 7:481–485

Haynes JR, Downing JR (1988) A recessive cellular mutation in v-fes-transformed mink cells restores contact inhibition and anchorage-dependent growth. Mol Cell Biol 8:2419–2427

Heidaran MA, Molloy CJ, Pangelinan M, Choudhury GG, Wang L, Fleming TP, Sakaguchi AY, Pierce JH (1992) Activation of the colony-stimulating factor 1 receptor leads to the rapid tyrosine phosphorylation of GTPase-activating protein and activation of cellular p21ras. Oncogene 7:147–152

Higgins PJ, Ryan MP (1989) Biochemical localization of the transformation-sensitive 52 kDA (p52) protein to the substratum contact regions of cultured rat fibroblasts. Biochem J 257:173–182

Higgins PJ, Ryan MP (1991) p52 (PAI-1) and actin expression in butyrate-induced flat revertants of v-ras-transformed rat kidney cells. Biochem J 279:883–890

Hirakawa T, Ruley HE (1988) Rescue of cells from ras oncogene-induced growth arrest by a second, complementing, oncogene. Proc Natl Acad Sci USA 85:1519–1523

Hoemann CD, Zarbl H (1990) Use of revertant cell lines to identify targets of v-fos transformation-specific alterations in gene expression. Cell Growth Diff 1:581–590

Hollstein M, Sidransky D, Vogelstein B, Harris CC (1991) p53 mutations in human cancers. Science 253:49–53

Hong HJ, Hsu L-C, Gould MN (1990) Molecular cloning of rat Krev-1 cDNA and analysis of the mRNA levels in normal and NMU-induced mammary carcinomas. Carcinogenesis 11:1245–1247

Hsu L-C, Gould MN (1991) Molecular cloning of Copenhagen rat Krev-1 and rap1B cDNAs and study of their association with mammary tumor resistance in the Copenhagen rat. Carcinogenesis 12:533–536

Huang DCS, Marshall CJ, Hancock JF (1993) Plasma membrane-targeted ras GTPase-activating protein is a potent suppressor of p21ras function. Mol Cell Biol 13:2420–2431

Huang S, Axelrod DE (1991) Altered post-translational processing of p21ras oncoprotein in a transformation-suppressed cell line. Oncogene 6:1211–1218

Hurlin PJ, Maher VM, McCormick JJ (1989) Malignant transformation of human fibroblasts caused by expression of a transfected T24 HRAS oncogene. Proc Natl Acad Sci USA 86:187–191

Ichikawa T, Ichikawa Y, Isaacs JT (1992) Genetic factors and suppression of metastatic ability of v-Ha-ras-transfected rat mammary cancer cells. Proc Natl Acad Sci USA 89:1607–1610

Itoh O, Kuroiwa S, Atsumi S, Umezawa K, Takeuchi T, Hori M (1989) Induction by the guanosine analogue oxanosine of reversion toward the normal phenotype of K-ras transformed rat kidney cells. Cancer Res 49:996–1000

Jelinek MA, Hassell JA (1992) Reversion of middle T antigen-transformed rat-2 cells by Krev-1: Implications for the role of p21c-ras in polyomavirus-mediated transformation. Oncogene 7:1687–1698

Jimenez B, Pizon V, Lerosey I, Bèranger F, Tavitian A, deGunzburg J (1991) Effects of the ras-related rap2 protein on cellular proliferation. Int J Cancer 49:471–479

Kaibuchi K, Mizuno T, Fuijoka H, Yamamoto T, Kishi K, Fukumoto Y, Hori Y, Takai Y (1991) Molecular cloning of the cDNA for stimulatory GDP/GTP exchange protein for smg p21s (ras-like small GTP-binding proteins) and characterization of stimulatory GDP/GTP exchange protein. Mol Cell Biol 11:2873–2880

Kaneko M, Horikoshi J (1989) Reversible suppression by nalidixic acid of anchorage-independent growth of mouse cells transformed by 3-methylcholanthrene or an activated c-Ha-ras gene. Br J Cancer 60:880–886

Katz E, Carter BJ (1986) A mutant cell line derived from NIH/3T3 cells: Two oncogenes required for in vitro transformation. J Natl Cancer Inst 77:909–914

Katz E, Samid D (1989) Transfection of EK-3, a subline of NIH 3T3, with the oncogene Ha-ras does not abolish its anchorage dependence. Europ J Cell Biol 49:221–224

Kawata M, Matsui Y, Kondo J, Hishida T, Teranishi Y, Takai Y (1988) A novel small molecular weight GTP-binding protein with the same putative effector domain as the ras proteins in bovine brain membranes. J Biol Chem 263:18965–18971

Kawata M, Kikuchi A, Hoshijima M, Yamamoto K, Hashimoto E, Yamamura H, Takai Y (1989a) Phosphorylation of smg p21, a ras p21-like GTP-binding protein, by cyclic AMP-dependent protein kinase in a cell-free system and in response to prostaglandin E1 in intact human platelets. J Biol Chem 264:15688–15695

Kawata M, Kikuchi A, Hoshijima M, Yamamoto K, Hashimoto E, Yamamura H, Takai Y (1989b) Phosphorylation of smg p21, a ras p21-like GTP-binding protein, by cyclic AMP-dependent protein kinase in a cell-free system and in response to prostaglandin E1 in intact human platelets. J Biol Chem 264:15688–15695

Kawata M, Kawahara Y, Sunako M, Araki S, Koide M, Tsuda T, Fukuzaki H, Takai Y (1991) The molecular heterogeneity of the smg-21/Krev-1/rap1 proteins, a GTP-Binding protein having the same effector domain as ras p21s, in bovine aortic smooth muscle membranes. Oncogene 6:841–848

Kenyon K, Contente S, Trackman PC, Tang J, Kagan HM, Friedman RM (1991) Lysyl oxidase and rrg messenger RNA. Science 253:802

Kikuchi A, Sasaki T, Araki S, Hata Y, Takai Y (1989) Purification and characterization from bovine brain cytosol of two GTPase-activating proteins specific for smg p21, a GTP-binding protein having the same effector domains as c-ras p21s. J Biol Chem 264:9133–9136

Kikuchi A, Kaibuchi K, Hori Y, Nonaka H, Sakoda T, Kawamura M, Mizuno T, Takai Y (1992) Molecular cloning of the human cDNA for a stimulatory GDP/GTP exchange protein for c-Ki-ras p21 and smg p21. Oncogene 7:289–293

Kim S, Mizoguchi A, Kikuchi A, Takai Y (1990) Tissue and subcellular distributions of the smg-21/rap1/Krev-1 proteins which are partly distinct from those of c-ras p21s. Mol Cell Biol 10:2645–2652

Kinzler KW, Nilbert MC, Su L-K, Vogelstein B, Bryan TM, Levy DB, Smith KJ, Preisinger AC, Hedge P, McKechnie D, Finniear R, Markham A, Groffen J, Boguski MS, Altschul SF, Horii A, Ando H, Miyoshi Y, Miki Y, Nishisho I, Nakamura Y (1991) Identification of FAP locus genes from chromosome 5q21. Science 253:661–664

Kitayama H, Sugimoto Y, Matsuzaki T, Ikawa Y, Noda M (1989) A ras-related gene with transformation suppressor activity. Cell 56:77–84

Kitayama H, Matsuzaki T, Ikawa Y, Noda M (1990) Genetic analysis of the Kirsten-ras-revertant 1 gene: Potentiation of its tumor suppressor activity by specific point mutations. Proc Natl Acad Sci USA 87:4284–4288

Kizaka S, Hakura A (1989) A cell mutant that exhibits temperature-dependent sensitivity to transformation by various oncogenes. Mol Cell Biol 9:5669–5675

Koi M, Johnson LA, Kalikin LM, Little PFR, Nakamura Y, Feinberg AP (1993) Tumor cell growth arrest caused by subchromosomal transferable DNA fragments from chromosome 11. Science 260:361–364

Koizumi M, Kamiya H, Ohtsuka E (1992) Ribozymes designed to inhibit transformation of NIH3T3 cells by the activated c-Ha-ras gene. Gene 117:179–184

Kollias G, Evens DJ, Ritter M, Beech J, Morris R, Grosveldt F (1987) Ectopic expression of Thy-1 in the kidneys of transgenic mice induces functional and proliferative abnormalities. Cell 51:21–31

Korn LJ, Siebel CW, McCormick F, Roth RA (1987) Ras p21 as a potential mediator of insulin action in Xenopus oocytes. Science 236:840–843

Kotani K, Kikuchi A, Doi K, Kishida S, Sakoda T, Kishi K, Takai Y (1992) The functional domain of the stimulatory GDP/GTP exchange protein (smg GDS) which interacts with the C-terminal geranylgeranylated region of rap-1/Krev-1/smg-p21. Oncogene 7:1699–1704

Krauss RS, Guadagno SN, Weinstein IB (1992) Novel revertants of H-ras oncogene-transformed R6-PKC3 cells. Mol Cell Biol 12:3117–3129

Krizman DB, Giovanella BC, Tainsky MA (1990) Susceptibility for N-ras-mediated transformation requires loss of tumor suppressor activity. Somat Cell Mol Genet 16:15–27

Krzyzosiak WJ, Shindookada N, Teshima H, Nakajima K, Nishimura S (1992) Isolation of genes specifically expressed in flat revertant cells derived from activated ras-transformed NIH- 3T3 cells by treatment with azatyrosine. Proc Natl Acad Sci USA 89:4879–4883

Kulesh DA, Greene JJ (1986) Shape-dependent regulation of proliferation in normal and malignant human cells and its alteration by interferon. Cancer Res 46:2793–2797

Kuzumaki N, Ogiso Y, Oda A, Fujita H, Suzuki H, Sato C, Müllauer L (1989) Resistance to oncogenic transformation in revertant R1 of human ras-transformed NIH 3T3 cells. Mol Cell Biol 9:2258–2263

Kyprianou N, Taylor-Papadimitriou J (1992) Isolation of azatyrosine-induced revertants from ras-transformed human mammary epithelial cells. Oncogene 7:57–63

Land H, Parada LF, Weinberg RA (1983) Tumorigenic conversion of primary embryo fibroblasts requires at least two cooperating oncogenes. Nature 304:596–602

Land H, Chen AC, Morgenstern JP, Parada LF, Weinberg RA (1986) Behavior of myc and ras oncogenes in transformation of rat embryo fibroblasts. Mol Cell Biol 6:1917–1925

Lane DP (1992) p53, Guardian of the genome. Nature 358:15–16

Lapetina EG, Lacal JC, Reep BR, yVedia LM (1989) A ras-related protein is phosphorylated and translocated by agonists that increase cAMP levels in human platelets. Proc Natl Acad Sci USA 86:3131–3134

Lazariskaratzas A, Smith MR, Frederickson RM, Jaramillo ML, Liu YL, Kung HF, Sonenberg N (1992) Ras mediates translation initiation factor-4E-induced malignant transformation. Gene Dev 6:1631–1642

Ledwith BJ, Manam S, Kraynack AR, Nichols WW, Bradley MO (1990) Antisense-fos RNA causes partial reversion of the transformed phenotypes induced by the c-Ha-ras oncogene. Mol Cell Biol 10:1545–1555

Lee SW, Tomasetto C, Sager R (1991) Positive selection of candidate tumor-suppressor genes by subtractive hybridization. Proc Natl Acad Sci USA 88:2825–2829

Legius E, Marchuk DA, Collins FS, Glover TW (1993) Somatic deletion of the neurofibromatosis type 1 gene in a neurofibrosarcoma supports a tumour suppressor gene hypothesis. Nature Genet 3:122–126

Lerosey I, Chardin P, deGunzburg J, Tavitian A (1991) The product of the rap2 gene, member of the ras superfamily. J Biol Chem 266:4315–4321

Lerosey I, Polakis P, Tavitian A, de Gunzburg J (1992) Regulation of the GTPase activity of the ras-related rap2 proptein. Biochem Biophys Res Comm 189:455–464

Levine AJ, Momand J, Finlay CA (1991) The p53 tumour suppressor gene. Nature 351:453–456

Li Y, Bollag G, Clark R, Stevens J, Conroy L, Fults D, Ward K, Friedman E, Samowitz W, Robertson M, Bradley P, McCormick F, White R, Cawthon R (1992) Somatic mutations in the neurofibromatosis-1 gene in human tumors. Cell 69:275–281

Low MG, Kincade PW (1985) Phosphatidylinositol is the membrane-anchoring domain of the Thy-1 glycoprotein. Nature 318:62–64

Lu X, Park SH, Thompson TC, Lane DP (1992) ras-induced hyperplasia occurs with mutation of p53, but activated ras and myc together can induce carcinoma without p53 mutation. Cell 70:153–161

Manfioletti G, Ruaro ME, Del Sal G, Philipson L, Schneider C (1990) A growth arrest-specific (gas) gene codes for a membrane protein. Mol Cell Biol 10:2924–2930

Marchuck DA, Saulino AM, Tavakkol R, Swaroop M, Wallace MR, Andersen LB, Mitchell AL, Gutmann DH, Boguski M, Collins FS (1991) cDNA cloning of the type 1 neurofibromatosis gene: complete sequence of the NF1 gene product. Genomics 11:931–940

Maridonneau-Parini I, de Gunzburg J (1992) Association of rap1 and rap2 proteins with the specific granules of human neutrophils. J Biol Chem 267:6396–6402

Marshall CJ (1993) Protein prenylation: a mediator of protein-protein interactions. Science 259:1865–1866

Marshall MS, Davis LJ, Keys RD, Mosser SD, Hill WS, Scolnick EM, Gibbs JB (1991) Identification of amino acid residues required for ras p21 target activation. Mol Cell Biol 11:3997–4004

Martin GA, Viskochil D, Bollag G, MyCabe PC, Crosier WJ, Haubruck H, Conroy L, Clark R, O'Connell P, Cawthon RM, Innis MA, McCormick F (1990) The GAP-related domain of the neurofibromatosis type 1 gene product interacts with ras p21. Cell 63:843–849

Martin GA, Yatani A, Clark R, Conroy L, Polakis P, Brown AM, McCormick F (1992) GAP domains responsible for ras p21-dependent inhibition of muscarinic atrial K+ channel currents. Science 255:192–194

Matsui Y, Kikuchi A, Kawata M, Kondo J, Teranishi Y, Takai Y (1990) Molecular cloning of smg p21B and identification of smg p21 purified from bovine brain and human platelets as smg p21B. Biochem Biophys Res Comm 166:1010–1016

Matsumoto M, Matsutani S, Sugita K, Yoshida H, Hayashi F, Terui Y, Nakai H, Uotani N, Kawamura Y, Matsumoto K, Shoji J, Yoshida T (1992) Depudecin: A novel compound inducing the flat phenotype of NIH3T3 cells doubly transformed by ras-oncogene and src-oncogene, produced by Alternaria-Brassicicola. J Antibiot 45:879–885

McCormick F (1989) ras GTPase activating protein: Signal transmitter and signal terminator. Cell 56:5-8

McCormick F (1993) How receptors turn Ras on. Nature 363:15–16

Medema RH, Wubbolts R, Bos JL (1991) Two dominant inhbitory mutants of p21ras interfere with insulin-induced gene expression. Mol Cell Biol 11:5963–5967

Michalovitz D, Halevy O, Oren M (1991) p53 mutations: Gains or losses? J Cell Biochem 45:22–29

Mitelman F, Heim S (1990) Chromosome abnormalities in cancer. Cancer Detect Prevention 14:527–537

Mizuno T, Kaibuchi K, Yamamoto T, Kawamura M, Sakoda T, Fujioka H, Matsuura Y, Takai Y (1991) A stimulatory GDP/GTP-exchange protein for smg p21 is active on the posttranslationally processed form of c-Ki-ras p21 and rho A p21. Proc Natl Acad Sci USA 88:6442–6446

Molloy CJ, Bottaro DP, Fleming TP, Marshall MS, Gibbs JB, Aaronson SA (1989) PDGF induction of tyrosine phosphorylation of GTPase activating protein. Nature 342:711–714

Molloy CJ, Fleming TP, Bottaro DP, Cuadrado A, Aaronson SA (1992) Platelet-derived growth factor stimulation of GTPase-activating protein tyrosine phosphorylation in control and c-H-ras-expressing NIH 3T3 cells correlates with p21(ras) activation. Mol Cell Biol 12:3903–3909

Monia BP, Johnston JF, Ecker DJ, Zounes MA, Lima WF, Freier SM (1992) Selective inhibition of mutant Ha-ras messenger RNA expression by antisense oligonucleotides. J Biol Chem 267:19954–19962

Morris A, Clegg C, Jones J, Rodgers B, Avery RJ (1980) The isolation and characterization of a clonally related series of murine retrovirus-infected mouse cells. J Gen Virol 49:105–113

Morris RJ (1985) Thy-1 in developing nervous tissue. Dev Neurosci 7:133–160

Mulcahy LS, Smith MR, Stacey DW (1985) Requirement for ras proto-oncogene function during serum-stimulated growth of NIH 3T3 cells. Nature 313:241–243

Mulder KM, Morris SL (1992) Activation of p21ras by transforming growth factor-beta in epithelial cells. J Biol Chem 267:5029–5031

Muleris M, Delattre O, Olschwang S, Dutrillaux A, Remvikos Y, Salmon R, Thomas G, Dutrillaux B (1990) Cytogenetic and molecular approaches of polyploidization in colorectal adenocarcinomas. Cancer Genet Cytogenet 44:107–118

Müllauer L, Suzuki H, Fujita H, Katabami M, Kuzumaki N (1991) Identification of genes that exhibit increased expression after flat reversion of NIH/3T3 cells transformed by human activated Ha-ras oncogene. Cancer Lett 59:37–43

Nakayasu M, Shima H, Aonuma S, Nakagama H, Nagao M, Sugimura T (1988) Deletion of transfected oncogenes from NIH 3T3 transformants by inhibitors of poly(ADP-ribose) polymerase. Proc Natl Acad Sci USA 85:9066–9070

Narayanan R, Lawlor KG, Schaapveld RQJ, Cho KR, Vogelstein B, Tran PBV, Osborne MP, Telang NT (1992) Antisense RNA to the putative tumor-suppressor gene DCC transforms rat-1 fibroblasts. Oncogene 7:553–561

Niles RM, Wilhelm SA, Thomas P, Zamcheck N (1988) The effect of sodium butyrate and retinoic acid on growth and CEA production in a series of human colorectal tumor cell lines representing different states of differentiation. Cancer Invest 6:39–45

Nishi T, Lee PSY, Oka K, Levin VA, Tanase S, Morino Y, Saya H (1991) Differential expression of two types of the neurofibromatosis type 1 (NF1) gene transcripts related to neuronal differentiation. Oncogene 6:1555–1559

Noda M, Selinger Z, Scolnick EM, Bassin RH (1983) Flat revertants isolated from Kirsten sarcoma virus-transformed cells are resistant to the action of specific oncogenes. Proc Natl Acad Sci USA 80:5602–5606

Noda M, Ko M, Ogura A, Liu D, Amano T, Takano T, Ikawa Y (1985) Sarcoma viruses carrying ras oncogenes induce differentiation-associated properties in a neuronal cell line. Nature 318:73–75

Noda M, Kitayama H, Matsuzaki T, Sugimoto Y, Okayama H, Bassin RH, Ikawa Y (1989) Detection of genes with a potential for suppressing the transformed phenotype associated with activated ras genes. Proc Natl Acad Sci USA 86:162–166

Noda M (1990) Expression cloning of tumor suppressor genes: a guide for optimists. Mol Carcinog 3:251–253

Nomura T, Ryoyama K, Okada G, Matano S, Nakamura S, Kameyama T (1992) Non-transformed, but not ras/myc-transformed, serum-free mouse embryo cells recover from growth suppression by azatyrosine. Jpn J Cancer Res 83:851–858

Nori M, Vogel US, Gibbs JB, Weber MJ (1991) Inhibition of v-src-induced transformation by a GTPase-activating protein. Mol Cell Biol 11:2812–2818

Norton JD, Cook F, Roberts PC, Clewley JP, Avery RJ (1984) Expression of Kirsten murine sarcoma virus in transformed nonproducer and revertant NIH/3T3 cells: Evidence for cell-mediated resistance to a viral oncogene in phenotypic reversion. J Virol 50:439–444

Nur-E-Kamal MSA, Sizeland A, D'Abaco G, Maruta H (1992) Aparagine 26, glutamic acid 31, valine 45, and tyrosine 64 of ras proteins are required for their oncogenicity. J Biol Chem 267:1415–1418

Nur-E-Kamal MSA,Varga M, Maruta H (1993) The GTPase-activating NF1 fragment of 91 amino acids reverses v-Ha-ras-induced malignant phenotype.J Biol Chem 268: 22331–22337

O'Connell P, Leach R, Cawthon RM, Culver M, Stevens J, Viskochil D, Fournier REK, Rich DC, Ledbetter DH, White R (1989) Two NF1 translocations map within a 600-kilobase segment of 17q11.2. Science 244:1087–1090

Ohmori T, Kikuchi A, Yamamoto K, Kim S, Takai Y (1989) Small molecular weight GTP-binding proteins in human platelet membranes. J Biol Chem 264:1877–1881

Ohmstede C-A, Farrell FX, Reep BR, Clemetson KJ, Lapetina EG (1990) RAP2B: A RAS-related GTP-binding protein from platelets. Proc Natl Acad Sci USA 87:6527–6531

Oshimura M, Gilmer TM, Barrett JC (1985) Nonrandom loss of chromosome 15 in Syrian hamster tumours induced by the v-Ha-ras plus v-myc oncogenes. Nature 316:636–639

Ozanne B, Vogel A (1974) Selection of revertants of Kirsten sarcoma virus transformed nonproducer BALB/3T3 cells. J Virol 14:239–248

Pan J, Roskelley CD, Luu-The V, Rojiani M, Auersperg N (1992) Reversal of divergent differentiation by ras oncogene-mediated transformation. Cancer Res 52:4269–4272

Paterson H, Reeves B, Brown R, Hall A, Furth M, Bos J, Jones P, Marshall C (1987) Activated N-ras controls the transformed phenotype of HT1080 human fibrosarcoma cells. Cell 51:803–812

Paterson HF, Self AJ, Garrett MD, Just I, Aktories K, Hall A (1990) Microinjection of recombinant p21rho induces rapid changes in cell morphology. J Cell Biochem 111:1001–1007

Pattengale PK, Stewart TA, Leder A, Sinn E, Muller W, Tepler I, Schmidt E, Leder P (1989) Animal models of human disease. Pathology and molecular biology of spontaneous neoplasms occuring in transgenic mice carrying and expressing activated cellular oncogenes. Am J Pathol 135:39–61

Peltomäki P, Aaltonen LA, Sistonen P, Pylkkänen L, Mecklin J-P, Järvinen H, Green JS, Jass JR, Weber JL, Leach FS, Petersen GM, Hamilton SR, De la Chapelle A, Vogelstein B (1993) Genetic mapping of a locus predisposing to human colorectal cancer. Science 260:810–812

Pizon V, Lerosey I, Chardin P, Tavitian A (1988a) Nucleotide sequence of a human cDNA encoding a ras-related protein (rap1B). Nucleic Acids Res 16:7719

Pizon V, Chardin P, Lerosey I, Olofsson B, Tavitian A (1988b) Human cDNAs rap 1 and rap 2 homologous to the Drosophila gene Dras 3 encode proteins closely related to ras in the "effector" region. Oncogene 3:201–204

Polakis PG, Rubinfeld B, Evans T, McCormick F (1991) Purification of a plasma membrane-associated GTPase-activating protein specific for rap1/Krev-1 from HL 60 cells. Proc Natl Acad Sci USA 88:239–243

Ponder B (1990) Neurofibromatosis gene cloned. Nature 346:703–704

Pratt CI, Wu SQ, Bhattacharya M, Kao CH, Gilchrist KW, Reznikoff CA (1992) Chromosome losses in tumorigenic revertants of EJ/ras- expressing somatic cell hybrids. Cancer Genet Cytogenet 59:180–190

Quilliam LA, Der CJ, Clark R, O'Rourke EC, Zhang K, McCormick FP, Bokoch GM (1990) Biochemical characterization of Baculovirus-expressed rap1A/Krev-1 and its regulation by GTPase-activating protein. Mol Cell Biol 10:2901–2908

Quilliam LA, Mueller H, Bohl BP, Prossnitz V, Sklar LA, Der CJ, Bokoch GM (1991) Rap1A is a substrate for cyclic AMP dependent protein kinase in human neutrophils. J Immunol 147:1628–1635

Quinn MT, Parkos CA, Walker L, Orkin SH, Dinauer MC, Jesaitis AJ (1989) Association of a Ras-related protein with cytochrome b of human neutrophils. Nature 342:198–200

Quinn MT, Mullen ML, Jesaitis AJ, Linner JG (1992) Subcellular distribution of the Rap1A protein in human neutrophils: colocalization and cotranslocation with cytochrome b559. Blood 79:1563–1573

Reddy EP, Reynolds RK, Santos E, Barbacid M (1982) A point mutation is responsible for the acquisition of transforming properties by the T24 human bladder carcinoma oncogene. Nature 300:149–152

Ridley AJ, Paterson HF, Noble M, Land H (1988) ras-mediated cell cycle arrest is altered by nuclear oncogenes to induce Schwann cell transformation. EMBO J 7:1635–1645

Ridley AJ, Hall A (1992) The small GTP-binding protein rho regulates the assembly of focal adhesions and actin stress fibers in response to growth factors. Cell 70:389–399

Roberts TM (1992) A signal chain of events. Nature 360:534–535

Rousseau-Merck MF, Pizon V, Tavitian A, Berger R (1990) Chromosome mapping of the human RAS-related RAP1A, RAP1B, and RAP2 genes to chromosomes 1p12-13, 12q14, and 13q34, respectively. Cytogenet Cell Genet 53:2–4

Rubinfeld B, Munemitsu S, Clark R, Conroy L, Watt K, Crosier WJ, McCormick F, Polakis P (1991) Molecular cloning of a GTPase activating protein specific for the Krev-1 protein p21rap1. Cell 65:1033–1042

Rubinfeld B, Crosier WJ, Albert I, Conroy L, Clark R, McCormick F, Polakis P (1992) Localization of the rap1GAP catalytic domain and sites of phosphorylation by mutational analysis. Mol Cell Biol 12:4634–4642

Ruch RJ, Madhukar BV, Trosko JE, Klaunig JE (1993) Reversal of ras-induced inhibition of gap-junctional intercellular communication, transformation, and tumorigenesis by lovastatin. Mol Carcinogen 7:50–59

Ruley HE (1983) Adenovirus early region 1A enables viral and cellular transforming genes to transform primary cells in culture. Nature 304:602–606

Ryan MP, Higgins PJ (1988) Cytoarchitecture of Kirsten sarcoma virus-transformed rat kidney fibroblasts: Butyrate-induced reorganization within the actin microfilament network. J Cell Physiol 137:25–34

Ryan MP, Higgins PJ (1989) Sodium-n-butyrate induces secretion and substrate accumulation of p52 in Kirsten sarcoma virus-transformed rat kidney fibroblasts. Int J Biochem 21:31–37

Sager R, Tanaka K, Lau CC, Ebina Y, Anisowicz A (1983) Resistance of human cells to tumorigenesis induced by cloned transforming genes. Proc Natl Acad Sci USA 80:7601–7605

Sakoda T, Kaibuchi K, Kishi K, Kishida S, Doi K, Hoshino M, Hattori S, Takai Y (1992) smg/rap-1/Krev-1 p21s inhibit the signal pathway to the c-fos promoter/enhancer from c-Ki-ras p21 but not from c-raf-1 kinase in NIH3T3-cells. Oncogene 7:1705–1711

Samid D, Flessate DM, Friedman RM (1987) Interferon-induced revertants of ras-transformed cells: Resistance to transformation by specific oncogenes and retransformation by 5-azacytidine. Mol Cell Biol 7:2196–2200

Santos E, Martin-Zanca D, Reddy EP, Pierotti MA, Della Porta G, Barbacid M (1984) Malignant activation of a K-ras oncogene in lung carcinoma but not in normal tissue of the same patient. Science 223:661–664

Satoh T, Endo M, Nakafuku M, Nakamura S, Kaziro Y (1990) Platelet-derived growth factor stimulates formation of active p21ras.GTP complex in Swiss mouse 3T3. Proc Natl Acad Sci USA 87:5993–5997

Satoh T, Nakafuku M, Miyajima A, Kaziro Y (1991) Involvement of ras p21 protein in signal transduction pathways from interleukin 2, interleukin 3, and granulocyte/macrophage colony-stimulating factor, but not from interleukin 4. Proc Natl Acad Sci USA 88:3314–3318

Satoh T, Nakafuku M, Kaziro Y (1992) Function of Ras as a molecular switch in signal transduction. J Biol Chem 267:24149–24152

Schäfer R, Iyer J, Iten E, Nirkko AC (1988) Partial reversion of the taransformed phenotype in HRAS-trasfered tumorigenic cells by transfer of a human gene. Proc Natl Acad Scie USA 85: 1590–1594

Schafer WR, Kim R, Sterne R, Thorner J, Kim S-H, Rine J (1989) Genetic and pharmacological suppression of oncogenic mutations in RAS genes of yeast and humans. Science 245:379–385

Schafer WR, Rine J (1992) Protein Prenylation: Genes, Enzymes, Targets, and Functions. Annu Rev Genet 30:209–237

Schneider C, King RM, Philipson L (1988) Genes specifically expressed at growth arrest of mammalian cells. Cell 54:787–793

Seizinger BR, Rouleau GA, Ozelius LJ, Lane AH, Faryniarz AG, Chao MV, Huson S, Korf BR, Parry DM, Pericak-Vance MA, Collins FS, Hobbs WJ, Falcone BG, Iannazzi JA, Roy JC, StGeorge-Hyslop PH, Tanzi RE, Bothwell MA, Upadyaya M, Harper P, Goldstein AE, Hoover DL, Bader JL, Spence MA, Mulvihill JJ, Aylsworth AS, Vance JM, Rossenwasser GOD, Gaskell PC, Roses AD, Martuza RL, Breakefield XO, Gusella JF (1987) Genetic linkage of von Recklinghausen neurofibromatosis to the nerve growth factor receptor gene. Cell 49:589–594

Seliger B, Pfizenmaier K, Schäfer R (1991) Short-time treatment with IFN-gamma induces stable reversion of ras transformed mouse fibroblasts. J Virol 65:6307–6311

Seremetis S, Inghirami G, Ferrero D, Newcomb EW, Knowles DM, Dotto G-P, Dalla-Favera R (1989) Transformation and plasmacytoid differentiation of EBV-infected human B lymphoblasts by ras oncogenes. Science 243:660–663

Sharma S, Schwarte-Waldhoff I, Oberhuber H, Schäfer R (1993) Functional activity of wild-type and mutant p53 transfected into human tumor cell lines carrying activated ras genes. Cell Growth Diff 4: 861–869

Sharma SV (1992) Melittin resistance: a counterselection for ras transformation. Oncogene 7:193–201

Shen WPV, Aldrich TH, Venta-Perez G, Franza BR, Furth ME (1987) Expression of normal and mutant ras proteins in human acute leukemia. Oncogene 1:157–165

Shier WT (1979) Activation of high levels of endogenous phospholipase A2 in cultured cells. Proc Natl Acad Sci USA 76:195–199

Shih C, Shilo BZ, Goldfarb MP, Dannenberg A, Weinberg RA (1979) Passage of phenotypes of chemically transformed cells via transfection of DNA and chromatin. Proc Natl Acad Sci USA 76:5714–5718

Shindo-Okada N, Makabe O, Nagahara H, Nishimura S (1989) Permanent conversion of mouse and human cells transformed by activated ras or raf genes to apparently normal cells by treatment with the antibiotic azatyrosine. Mol Carcinog 2:159–167

Shirasawa S, Furuse M, Yokoyama N, Sasazuki T (1993) Altered growth of human colon cancer cell lines disrupted at activated Ki-ras. Science 260:85–88

Simon MA, Bowtell DDL, Dodson GS, Laverty TR, Rubin GM (1991) Ras1 and a putative guanine nucleotide exchange factor perform crucial steps in signaling by the sevenless protein tyrosine kinase. Cell 67:701–716

Smith MR, DeGudicibus SJ, Stacey DW (1986) Requirement for c-ras proteins during viral oncogene transformation. Nature 320:540–543

Spandidos DA, Wilkie NM (1984) Malignant transformation of early passage rodent cells by a single mutated human oncogene. Nature 310:469–475

Spandidos DA, Wilkie NW (1988) The normal human H-ras1 gene can act as an onco-suppressor. Br J Cancer 58,Suppl.IX:67–71

Stanbridge EJ (1991) Human tumor suppressor genes. Annu Rev Genet 24:615–657

Stephenson JR, Reynolds RK, Aaronson SA (1973) Characterization of morphologic revertants of murine and avian sarcoma virus-transformed cells. J Virol 11:218–222

Stern PL (1973) Thy-1 alloantigen on mouse and rat fibroblasts. Nature 246:76–78

Stevenson M, Volsky DJ (1986) Activated v-myc and v-ras oncogenes do not transform normal human lymphocytes. Mol Cell Biol 6:3410–3417

Stoddart JH, Lane MA, Niles RM (1989) Sodium butyrate suppresses the transforming activity of an activated N-ras oncogene in human colon carcinoma cells. Exp Cell Res 184:16–27

Sugimoto Y, Ikawa Y, Nakauchi H (1991) Thy-1 as a negative growth regulator in ras-transformed mouse fibroblasts. Cancer Res 51:99–104

Sukumar S, Notario V, Martin-Zanca D, Barbacid M (1983) Induction of mammary carcinomas in rats by nitroso-methylurea involves malignant activation of H-ras-1 locus by single point mutations. Nature 306:658–661

Suzuki H, Takahashi K, Kubota Y, Shibahara S (1992) Molecular cloning of a cDNA coding for neurofibromatosis type-1 protein isoform lacking the domain related to ras GTPase-activating protein. Biochem Biophys Res Commun 187:984–990

Suzuki N, Choe H-R, Nishida Y, Yamawaki-Kataoka Y, Ohnishi S, Tamaoki T, Kataoka T (1990) Leucine-rich repeats and carboxyl terminus are required for interaction of yeast adenylate cyclase with RAS proteins. Proc Natl Acad Sci USA 87:8711–8715

Suzuki Y, Suzuki H, Kayama T, Yoshimoto T, Shibahara S (1991) Brain tumors predominantly express the neurofibromatosis type 1 gene transcripts containing the 63 base insert in the region coding for GTPase activating protein-related domain. Biochem Biophys Res Comm 181:955–961

Talbot N, Tagliaferri P, Yanagihara K, Rhim JS, Bassin RH, Benade LE (1988) A ph-dependent differential cytotoxicity of ouabain for human cells transformed by certain oncogenes. Oncogene 3:23–26

Tanaka K, Matsumoto K, Toh-e A (1989) IRA1, an inhibitory regulator of the RAS-cyclic AMP pathway in Saccharomyces cerevisiae. Mol Cell Biol 9:757–768

Tanaka K, Nakafuku M, Tamanoi F, Kaziro Y, Matsumoto K, Toh-e A (1990a) IRA2, a second gene of Saccharomyces cerevisiae that encodes a protein with a domain homologous to mammalian ras GTPase-activating protein. Mol Cell Biol 10:4303–4313

Tanaka K, Nakafuku M, Satoh T, Marshall MS, Gibbs JB, Matsumoto K, Kaziro Y, Toh-e A (1990b) S. cerevisiae genes IRA1 and IRA2 encode proteins that may be functionally equivalent to mammalian ras GTPase activating protein. Cell 60:803–807

Tanaka K, Lin BK, Wood DR, Tamanoi F (1991) IRA2, an upstream negative regulator of RAS in yeast, is a RAS GTPase-activating protein. Proc Natl Acad Sci USA 88:468472

Teinturier C, Danglot G, Slim R, Pruliere D, Launay JM, Bernheim A (1992) The neurofibromatosis-1 gene transcripts expressed in peripheral nerve and neurofibromas bear the additional exon located in the GAP domain. Biochem Biophys Res Commun 188:851–857

The I, Murthy AE, Hannigan GE, Jacoby LB, Menon AG, Gusella JF, Bernards A (1993) Neurofibromatosis type 1 gene mutations in neuroblastoma. Nature Genet 3:62–66

Thompson TC, Southgate J, Kitchener G, Land H (1989) Multistage carcinogenesis induced by ras and myc oncogenes in a reconstituted organ. Cell 56:917–930

Torti M, Marti KB, Altschuler D, Yamamoto K, Lapetina EG (1992) Erythropoietin induces p21ras activation and p120GAP tyrosine phosphorylation in human erythroleukemia cells. J Biol Chem 267:8293–8298

Trackman PC, Pratt AM, Wolanski A, Tang S-S, Offner GD, Troxler RF, Kagan HM (1990) Cloning of rat aorta lysyl oxidase cDNA: Complete codons and predicted amino acid sequence. Biochemistry 29:4863–4870

Trahey M, McCormick F (1987) A cytoplasmic protein stimulates normal N-ras p21 GTPase, but does not affect oncogenic mutants. Science 238:542–545

Trahey M, Wong G, Halenbeck R, Rubinfeld B, Martin GA, Ladner M, Long CM, Crosier WJ, Watt K, Koths K, McCormick F (1988) Molecular cloning of two types of GAP complementary DNA from human placenta. Science 242:1697–1700

Tsai M-H, Yu C-L, Stacey DW (1990) A cytoplasmic protein inhibits the GTPase activity of H-ras in a phospholipid-dependent manner. Science 250:982–985

Tse AGD, Barclay AN, Watts A, Williams AF (1985) A glycophospholipid tail at the carboxy terminus of the Thy-1 glycoprotein of neurons and thymocytes. Science 230:1003–1008

Tucker RF, Butterfield CE, Folkman J (1981) Interaction of serum and cell spreading affects the growth of neoplastic and nonneoplastic fibroblasts. J Supramol Struct Cell Biochem 15:29–40

Viskochil D, Buchberg AM, Xu G, Cawthon RM, Stevens J, Wolff RK, Culver M, Carey JC, Copeland NG, Jenkins NA, White R, O'Connell P (1990) Deletions and a translocation interrupt a cloned gene at the neurofibromatosis type 1 locus. Cell 62:187–192

Vogel A, Pollack R (1974) Isolation and characterization of revertant cell lines. VI. Susceptibility of revertants to retransformation by simian virus 40 and murine sarcoma virus. J Virol 14:1404–1410

Vogel US, Dixon RAF, Schaber MD, Diehl RE, Marshall MS, Scolnick EM, Sigal IS, Gibbs JB (1988) Cloning of bovine GAP and its interaction with oncogenic ras p21. Nature 335:90–93

Vogelstein B, Fearon ER, Hamilton SR, Kern SE, Preisinger AC, Leppert M, Nakamura Y, White R, Smits AMM, Bos JL (1988) Genetic alterations during colorectal-tumor development. N Engl J Med 319:526–532

Vogt M, Lesley J, Bogenberger J, Volkman S, Haas M (1986) Coinfection with viruses carrying the v-Ha-ras and v-myc oncogenes lead to growth factor independence by an indirect mechanism. Mol Cell Biol 6:3545–3549

Wallace MR, Marchuk DA, Andersen LB, Letcher R, Odeh HM, Saulino AM, Fountain JW, Brereton A, Nicholson J, Mitchell AL, Brownstein BH, Collins FS (1990) Type 1 neurofibromatosis gene: identification of a large transcript disrupted in three NF1 patients. Science 249:181–186

Wang S-Y, Bassin RH, Racker E (1988) Effect of high K+ hypertonicity and ouabain on MeAIB uptake and on growth of c-myc and v-ras transfected rat fibroblasts. Oncogene 3:53–57

Whitman M, Melton DA (1992) Involvement of p21ras in Xenopus mesoderm induction. Nature 357:252–254

Williams AF (1985) Immunoglobulin-related domains for cell surface recognition. Nature 314:579–580

Wilson DM, Yang D, Dillberger JE, Dietrich SE, Maher VM, McCormick JJ (1990) Malignant transformation of human fibroblasts by a transfected N-ras oncogene. Cancer Res 50:5587–5593

Winter E, Yamamoto F, Almoguera C, Perucho M (1985) A method to detect and characterize point mutations in transcribed genes: amplification and overexpression of the mutant c-Ki-ras allele in human tumors. Proc Natl Acad Sci USA 82:7575–7579

Wisdom R, Verma IM (1990) Revertants of v-fos-transformed rat fibroblasts: suppression of transformation is dominant. Mol Cell Biol 10:5626–5633

Xu G, Lin B, Tanaka K, Dunn D, Wod D, Gesteland R, White R, Weiss R, Tamanoi F (1990a) The catalytic domain of the neurofibromatosis type 1 gene product stimulates ras GTPase and complements ira mutants of S. cerevisiae. Cell 63:835–841

Xu G, O'Connell P, Viskochil D, Cawthon R, Robertson M, Culver M, Dunn D, Stevens J, Gesteland R, White R, Weiss R (1990b) The neurofibromatosis type 1 gene encodes a protein related to GAP. Cell 62:599–608

Yagle MK, Parruti G, Yu W, Ponder BAJ, Solomon E (1990) Genetic and physical map of the von Recklinghausen neurofibromatosis (NF1) region on chromosome 17. Proc Natl Acad Sci USA 87:7255–7259

Yamada H, Horikawa I, Hashiba H, Oshimura M (1990a) Normal human chromosome 1 carries suppressor activity for various phenotypes of a Kirsten murine sarcoma virus-transformed NIH/3T3 cell line. Jpn J Cancer Res 81:1095–1100

Yamada H, Omata-Yamada T, Wakabayashi-Ito N, Carter SG, Lengyel P (1990b) Isolation of recessive (mediator-) revertants from NIH 3T3 cells transformed with a c-Ha-ras oncogene. Mol Cell Biol 10:1822–1827

Yamamoto T, Kaibuchi K, Mizuno T, Hiroyoshi M, Shirataki H, Takai Y (1990) Purification and characterization from bovine brain cytosol of proteins that regulate the GDP/GTP exchange reaction of smg p21s, ras p21-like GTP-binding proteins. J Biol Chem 265:16626–16634

Yamasaki H (1991) Aberrant expression and function of gap junctions during carcinogenesis. Environ Health Perspect 93:191–197

Yanagihara K, Ciardiello F, Talbot N, McGready ML, Cooper H, Benade L, Salmon DS, Bassin RH (1990) Isolation of a new class of "flat" revertants from ras-transformed NIH3T3 cells using cis-4-hydroxy-L-proline. Oncogene 5:1179–1186

Yatani A, Okabe K, Polakis P, Halenbeck R, McCormick F, Brown AM (1990) Ras p21 and GAP inhibit coupling of muscarinic receptors to atrial K+ channels. Cell 61:769–776

Yatani A, Quilliam LA, Brown AM, Bokoch GM (1991) Rap1A antagonizes the ability of ras and ras-GAP to inhibit muscarinic K+ channels. J Biol Chem 266:22222–22226

Yoshida Y, Kawata M, Miura Y, Musha T, Sasaki T, Kikuchi A, Takai Y (1992) Microinjection of smg/rap1/Krev-1 p21 into Swiss 3T3-cells induces DNA synthesis and morphological changes. Mol Cell Biol 12:3407–3414

Young J, Searle J, Stitz R, Cowen A, Ward M, Chenevix-Trench G (1992) Loss of heterozygosity at the human RAP1A/Krev-1 locus is a rare event in colorectal tumors. Cancer Res 52:285–289

Zambetti GP, Olson D, Labow M, Levine AJ (1992) A mutant p53 protein is required for maintenance of the transformed phenotype in cells transformed with p53 plus ras cDNAs. Proc Natl Acad Sci USA 89:3952–3956

Zarbl H, Sukumar S, Arthur AV, Martin-Zanca D, Barbacid M (1985) Direct mutagenesis of Ha-ras-1 oncogenes by N-nitroso-N-methylurea during initiation of mammary carcinogenesis in rats. Nature 315:382–385

Zarbl H, Latreille J, Jolicoeur P (1987) Revertants of v-fos-transformed fibroblasts have mutations in cellular genes essential for transformation by other oncogenes. Cell 51:357-369

Zhang K, Noda M, Vass WC, Papageorge AG, Lowy DR (1990) Identification of small clusters of divergent amino acids that mediate the opposing effects of ras and Krev-1. Science 249:162–165

Zucker S, Lysik RM, Malik M, Bauer BA, Caamano J, Kleinszanto AJP (1992) Secretion of gelatinases and tissue inhibitors of metalloproteinases by human lung cancer cell lines and revertant cell lines: not an invariant correlation with metastasis. Int J Cancer 52:366–371

Editor-in-charge: Prof. H. Grunicke and Prof. H. Schweiger

Note added in proof

A number of recently published papers describing new revertant cell lines, genes associated with *ras* reversion, and inhibitors of *ras* expression or function could no longer be discussed in this article. Of particular interest is the finding that several genes are able to suppress a variety of the parameters of neoplastic transformation mediated by *ras*. These genes include a deletion mutant of the proto-oncogene *c-jun* (Brown et al. 1993) and the genes encoding the cognate heat shock protein hsc 70 (Yehiely and Oren 1992), tropomyosin-1 (Prasad et al. 1993) and the mutant form of gelsolin (Müllauer et al. 1993).

References

Brown PH, Alani R, Preis LH, Szabo E, Birrer MJ (1993) Suppression of Oncogene-Induced Transformation by a Deletion Mutant of c-jun. Oncogene 8:877–886

Müllauer L, Fujita H, Ishizaki A, Kuzumaki N (1993) Tumor-suppressive function of mutated gelsolin in ras-transformed cells. Oncogene 8:2531–2536

Prasad GL, Fuldner RA, Cooper HL (1993) Expression of Transduced Tropomyosin-1 cDNA Suppresses Neoplastic Growth of Cells Transformed by the ras Oncogene. Proc Natl Acad Sci USA 90:7039–7043

Yehiely F, Oren M (1992) The gene for the rat heat-shock cognate, hsc70, can suppress oncogene-mediated transformation. Cell Growth Differ 3:803–809

Rev. Physiol. Biochem. Pharmacol., Vol. 124
© Springer-Verlag 1994

Regulation of Gene Expression by Prolactin

W. DOPPLER

Contents

1 Introduction: Prolactin – A Multifunctional Regulator

Long before the polypeptide hormone prolactin was biochemically characterized, the existence of a lactogenic hormone was postulated. Extracts from
the anterior lobe of the pituitary stimulated the mammary gland to synthesize
milk protein genes (Stricker and Grueter 1928). Later on, a great number of

Institut für Medizinische Chemie und Biochemie der Universität Innsbruck, Fritz-Pregl-
Strasse 3, A-6020 Innsbruck, Austria

other physiological effects were reported to be triggered by the hormone. When Nicoll (1974) wrote his comprehensive review on the physiological actions of prolactin, the number of reported actions of prolactin already exceeded the number reported for all other adenophysial hormones combined. They include effects on the growth and differentiation of organs involved in reproduction, namely, ovary, testis (Dombrowicz et al. 1992), prostate (Pérez-Villamil et al. 1992), and mammary gland. Prolactin was also found to have an effect on the islets of the pancreas (Brelje et al. 1993) and has been described to be involved in osmoregulation (Leontic and Tyson 1977), in the initiation of behavioral effects and in the regulation of the immune system in mammals and birds (Gala 1991; Skwarlo-Sonta 1992). In development, it exerts a juvenilizing action in amphibian metamorphosis by counteracting the effects of thyroid hormone (Tata et al. 1991).

In the vertebrate pituitary, prolactin is synthesized in specialized cells, the lactotrophs. From the five cell types in the anterior pituitary, the lactotrophs appear last near birth in the developing gland. Estrogens induce the expansion of lactotrophs and stimulate the expression of the prolactin gene within these cells. Work on the ontogeny of this cell type has been reviewed recently (Voss and Rosenfeld 1992). Secretion of prolactin from the gland is subject to hypothalamic control and inhibited by dopamine. The secretion is pulsatile with a sexually dimorphic pattern (Arey et al. 1989). This might be important for a differential effect of the hormone on the gene expression in males vs. females as it has been described for the related growth hormone (Mode et al. 1989; Legraverend et al. 1992).

Several reports indicate that prolactin is also produced outside the pituitary in placenta (Hiroka et al. 1991), decidua (Di Mattia et al. 1990) and myometrium (Gellerson et al. 1991). The extra-pituitary produced prolactin encodes for the same size of protein synthesized in lactotrophs. However, the transcripts contain an additional 5'-noncoding exon. Prolactin-specific transcripts and immunoreactive protein were also found in human thymocytes and lymphoid cell lines (O'Neal et al. 1992; Pellegrini et al. 1992). The detection of prolactin transcripts in lymphocytes support the hypothesis that the hormone has also the properties of a cytokine. Since both placenta and thymocytes have been shown to express also prolactin receptors (PLR-R; Sect. 2.1), prolactin can act in an autocrine or paracrine manner there. Two reports describe the expression of prolactin transcripts in the mammary gland of rats (Steinmetz et al. 1993; Kurtz et al. 1993), suggesting that some of the prolactin secreted into the milk is produced directly in the mammary gland.

The hormone is subject to posttranslational modifications. Four differently glycosylated forms of prolactin have been described (Sinha et al. 1991). In the human pituitary, 10%–15% of prolactin is glycosylated. This modification appears to alter the biological activity of the hormone (Sinha et al. 1991; Markoff et al. 1988). A 16-kDa fragment of prolactin, but not the unprocessed prolactin has been found to specifically inhibit fibroblast growth factor stimulated growth of capillary endothelial cells (Ferrara et al. 1991). A receptor distinct from the PRL-R is postulated to mediate this response (Clapp and Weiner 1992).

In addition to prolactin variants, hormones have been identified which are structurally and functionally related to prolactin but are expressed from different genes. They form the prolactin family of hormones, which include placental lactogens, prolactin-like proteins, and proliferins (Soares et al. 1991). Placental trophoblasts are the major source of these mainly glycosylated peptides. There is a strong species difference in the expression pattern and structure of the various members of this family. Members of the proliferin family have been found only in mice to date. Placenta lactogen found in humans is closely related to growth hormone, whereas the placental lactogens found in nonprimate mammals are more related to prolactin than to growth hormone. In primates, growth hormone can also induce prolactin-like actions. This has been attributed to the unique property of primate growth hormone to bind to the PRL-R. Zinc ion, in complex with the human growth hormone has been described to mediate the high-affinity binding (Cunningham et al. 1990). A more than 1000-fold higher affinity of human growth hormone to a bacterial expressed extracellular domain of the human PRL-R was observed in the presence of zinc ions. In a later report (Rozakis-Adcock and Kelly 1991), high-affinity binding of human growth hormone without prior addition of $ZnCl_2$ was observed. However, the authors could not exclude the possibility that trace amounts of zinc were responsible for the effect. Experiments with transgenic animals harboring the gene for the human growth hormone revealed that human growth hormone is also mammotrophic and lactogenic in mice (Bchini et al. 1991). Again, binding of the human growth hormone to the mouse PRL-R appears to be responsible for that effect. The species-specific evolution of multiple hormones with prolactin-like actions is not surprising in view of the great number of physiological effects induced by prolactin. It is speculated that the diversity increases the fidelity of the complex stage and tissue-specific synthesis, secretion, and delivery of the hormone to the multiple target cells and that some hormones might also have slightly different effects when bound to their receptor (Soares et al. 1991).

In contrast to the abundant knowledge about physiological actions of prolactin and its relatives, progress in the understanding of biochemical mechanisms by which the hormone exerts its multiple specific functions was slow until the end of the last decade. Research was severely impeded by the failure to identify the target molecules of prolactin action in the cell. The situation has changed dramatically since the cloning of the PRL-R from rat liver in 1988 (Boutin et al. 1988). In addition, the identification of *cis*-acting sequences mediating the response to prolactin (Doppler et al. 1989) in the promoter of milk protein genes has set the stage to identify nuclear factors involved in the regulation of gene expression by prolactin. Subsequent work has established unforeseen and exciting relationships in the mechanisms by which prolactin or other extracellular signaling proteins influence gene expression.

2 The Prolactin Receptor: Member of a Large Family

2.1 Structure and Expression

A wealth of information about the structure of the PRL-Rs expressed in mammals was already available 3 years after cloning of the rat liver receptor. To date, the primary structure of PRL-Rs has been described from five mammalian species (Table 1). The receptors are organized into an extracellular region, a short transmembrane region, and an intracellular region of variable length. When expressed in COS cells, the cloned receptors bound prolactin and human growth hormone with approximately the same affinity $(2\text{–}3.10^{-10}\,M)$ previously described for purified receptor preparations. The hormones of the prolactin family also bind with high affinity. There is some evidence for the existence of a distinct receptor for placental lactogen (Freemark and Comer 1989) with distinct binding behavior when compared to the PRL-R. However, the cloning of such a receptor has not been described so far.

In all animal species studied, several transcripts were specifically hybridizing to PRL-R complementary DNA (cDNA)-derived probes (Table 1).

Table 1. Cloned forms of the prolactin receptor

Species	Transcript (kbp)	Protein size (amino acids)	Reference
Human	2.8, 3.5, 7.3, 2.7, 3.5, 8.6, 10.5	598	Boutin et al. 1989 Ormandy et al.1993
Cow	2.6, 3.8, 4.4	557	Scott et al. 1992
Rabbit	2.8, 4, 6.5, 10 2.7, 3.4, 6.2, 10.5	592	Edery et al. 1989 Dusanter-Fourt et al. 1991
Rat	1.8, 2.5, 3, 5.5 1.8, 2.1, 2.6, 4.6, 9.7 6.7, 10.1	291, 591	Shirota et al. 1990; Boutin et al. 1988 Hu and Dufau 1991 Dardenne et al. 1991
Mouse	1.4, 2.4, 3.5, 4.2, 8.3, 9, 10	292, 303, 310, 597	Davis and Linzer 1989a; Clark and Linzer 1993; Buck et al. 1992

They appear to result from differential splicing of one primary transcript (Kelly et al. 1991; Dusanter-Fourt et al. 1991). A final proof for this notion is expected to come from the publication of the exon/intron structure of a PRL-R gene. In humans, cows, and rabbits, the available cDNA sequences suggest that only one form of the PRL-R protein is expressed, with a size ranging from 557 amino acids in cows to 598 amino acids in humans. In rats, two receptor forms – a short form and a long form – were identified. These two receptors have identical extracellular domains but differ by 300 amino acids in the length of their intracellular domain. Mice exhibit the most complex pattern of PRL-R gene expression. Filter hybridization analysis of transcripts has revealed the existence of at least seven transcripts in mouse organs (Buck et al. 1992), coding for four different receptor proteins (Clarke and Linzer 1993; Table 1). Again, these four murine receptor forms have identical extracellular domains and differ only in the cytoplasmic part of the receptor. The prolactin binding proteins found in serum (Amit et al. 1992) and in milk (Postel-Vinay et al. 1991) might also be encoded by some of the PRL-R transcripts, as it has been described for the growth hormone-binding protein (Tiong and Herington 1992 and references therein).

Our present knowledge about the expression pattern of the PRL-R in animals and cell lines is mainly derived from experiments measuring receptor-specific transcripts or from the analysis of the specific binding sites for prolactin. The availability of suitable antibodies has limited so far a direct determination of receptor protein levels and sizes in tissues. Transcripts of the receptor were found in many organs (Kelly et al. 1991). Especially high levels were detected in the female liver, in the ovary, and in the mammary gland. Polymerase chain reaction techniques allowed detection of receptor transcripts in lymphoid tissues, lymphoid cell lines (O'Neal et al. et al. 1991; Pellegrini et al. 1992; Koh and Phillips 1993), and in the hypothalamus (Chiu et al. 1992). The latter results were confirmed by autoradiographic localization of prolactin binding sites (Crumeyrolle-Arias et al. 1993). PRL-Rs were also characterized in the human brain and the choroid plexus (Di Carlo et al. 1992). Monoclonal antibodies directed against the rat PRL-R allowed the identification of PRL-Rs in human and murine thymic epithelial cells (Dardenne et al. 1991). The antibodies were also able to modulate the function of these cells, presumably by activating the receptors.

The significance of the multiple forms of receptor protein expressed in rodents is not clear. The proteins only differ in their intracellular domain, and it has been suggested that this might result in differential coupling to intracellular signaling systems (Davis and Linzer 1989a). Evidence for that notion comes from cotransfection experiments of the long and the short form of the rat receptor with a milk protein gene promoter construct (Lesueur et al. 1991). Only the long form of the receptor was capable of mediating the effect of prolactin on the expression of the reporter gene linked to the promoter construct. The transcripts for the different receptor forms have been reported to be independently regulated during pregnancy and lactation in a tissue-specific manner (Jahn et al. 1991; Jolicoeur et al. 1989; Hu and Dufau 1991; Buck et al. 1992; Clarke and Linzer 1993). Of special interest is the regulation of receptor transcripts in the ovary during pregnancy: In that organ, prolactin has two well-described functions (Nicoll 1974): It acts as a luteotrophic hormone during pregnancy and thereby stimulates production of progesterone and synthesis of chorionic gonadotropin receptor (Gåfvels et al. 1992) in the corpus luteum. In addition, prolactin promotes as a luteolytic hormone the degeneration of corpora lutea from previous reproductive cycles. It is speculated that the organ serves also in relaying signals between prolactin and members of the prolactin family (Soares et al. 1991). In accordance with this diversity of functional roles, the ovaries are the organs with the most complex pattern of PRL-R transcripts (Hu and Dufau 1991; Buck et al. 1992). In situ hybridizations established a differential expression of the transcripts during pregnancy in granulosa cells and the corpus luteum (Clarke and Linzer 1993). Most interestingly, expression of one short receptor form was uniquely detected in atretic follicles at midgestation, thus implicating a role of this receptor form in mediating the luteolytic functions of prolactin.

The structure of the PRL-R has been compared extensively with published data obtained from other cloned receptors. The closest relative is the receptor for growth hormone (Leung et al. 1987). In addition, the receptors for a number of cytokines including the interferons have regions of relatively low but significant sequence homology with the PRL-R in their extracellular domains (Bazan 1989, 1990; Thoreau et al. 1991; Kelly et al. 1991). On the basis of these findings, the PRL-R has been grouped together with the growth hormone receptor (GH-R) into an expanded cytokine receptor superfamily.

2.2 Mechanism of Activation by Hormone

2.2.1 Site of Interaction with the Hormone

Regions in the extracellular domain of the receptor critically for the binding of the hormone have been mapped (Rozakis-Adcock and Kelly 1991). An N-terminal cystein-rich domain was found to be important for the specificity of binding. In addition, the WS×WS motif (tryptophan, serine, any amino acid, tryptophan, serine) of the receptor, which is conserved in the cytokine superfamily (Bazan 1990), is required for high-affinity binding and it has been proposed that this may serve as a target site for interaction with an accessory protein necessary for the formation of high-affinity binding sites (Rozakis-Adcock and Kelly 1991). The domains of the hormone required for binding have not been mapped yet. The availability of monoclonal antibodies directed against a number of distinct epitopes of prolactin (Staindl et al. 1987; Scammell et al. 1992) and the analysis of mutated prolactin molecules (Luck et al. 1992) should facilitate the localization. It is usually assumed that the receptor localized in the plasma membrane is contacted and activated by the hormone. There has also been a report on the requirement for nuclear expressed prolactin to stimulate interleukin-2 (IL-2)-dependent proliferation of T lymphocytes (Clevenger et al. 1991), implicating a second site of activation of the PRL-R or another unkown prolactin target in the cell. By contrast, in Nb2 rat lymphoma cells, which are believed to be thymocytes at an intermediate stage of differentiation (Gala 1991), intracellular prolactin had no effect on stimulation of cell proliferation (Davis and Linzer 1988).

2.2.2 Evidence for Receptor Dimerization

An important mechanism in the activation of receptors with tyrosine kinase activity is ligand-induced dimerization (Ullrich and Schlessinger 1990). There is now plenty of evidence that the same is true for a number of receptors belonging to the cytokine receptor superfamily (Stahl and Yancopoulos 1993). Cytokine receptors can be classified into four different subgroups by the subunit structure of the formed dimers (Fig. 1). In subgroup A, the ligand binds in a first step with one surface (site 1) to a receptor monomer. In a second step, dimerization is triggered by the contact of a

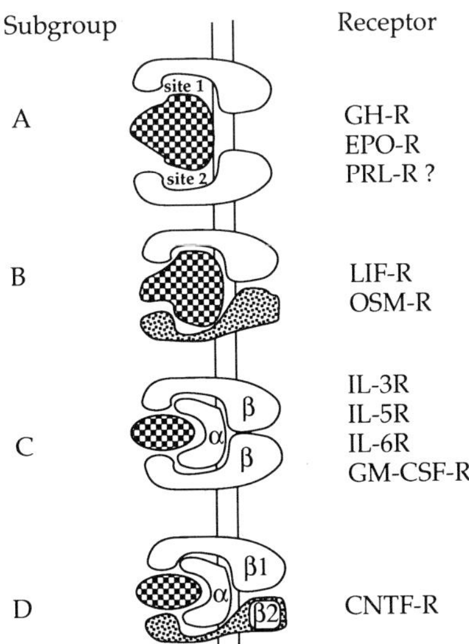

Fig. 1. Subunit structure of dimerized cytokine receptors. In all subgroups one ligand molecule triggers the dimerization of two receptor subunits. In subgroups *C* and *D*, an additional α component of the receptor is involved in the binding of the ligand and in the formation of the dimer. Receptor homodimers (subgroups *A* and *C*) or heterodimers (subgroups *B* and *D*) have been described. In subgroup *A*, the two contact surfaces (*site 1* and *site 2*) of the ligand with the receptors are indicated. For abbreviations see text

second assymetric surface (site 2) of the ligand to a second receptor molecule. The GH-R is the prototype for subgroup A. The crystal structure of two molecules of the GH-R extracellular binding domain complexed by one molecule of growth hormone (Ultsch et al. 1991; de Vos et al. 1992) has been obtained at 2.8-Å resolution, which allowed a precise mapping of the contact surfaces employed by the hormone and receptor molecules (reviewed by Demeyts 1992) and facilitates the rational design of receptor antagonists (Fuh et al. 1993). It is believed that the receptor for erythropoetin (EPO-R) also belongs to subgroup A. Subgroup B comprises the receptors for leukemia inhibitory factor (LIF-R) and oncostatin M (OSM-R). Here, heterodimers of two different receptor subunits, encoded by different genes, are formed by the ligand. In subgroups C and D, dimerization is dependent on an additional receptor subunit (α), which binds specifically to the ligand. Examples are the receptors for the IL-3, IL-5, and IL-6 (IL-3R, IL-5R, and IL-6R) and the granulocyte–macrophage colony-stimulating factor (GM-CSF-R), where the dimerized β-receptor components are homodimers (subgroup C). The ciliary neurotrophic factor receptor (CNTF-R) forms heterodimers of two different components (β1 and β2) in the presence of the cytokine complexed to the α-receptor subunit (subgroup D).

Indirect evidence obtained with Nb2 rat lymphoma cells suggests that the PRL-R belongs to subgroup A (Table 1): First, antibodies were able to activate the receptor presumably by induction of dimerization (Elberg et al. 1990). Second, human growth hormone mutants, with mutations in their binding sites for either the first or second GH-R molecule (first and second site mutants, see also schema for subgroup A receptors in Fig. 1), had functional effects expected for a situation in which PRL-R and the GH-R are dimerizing by the same mechanism: Both site 1 and site 2 mutants showed greatly reduced agonistic effects, whereas only site 2 mutants had antagonistic properties (Fuh et al. 1993). A mutant form of placental lactogen II, which was able to bind to the PRL-R, but was not active in the Nb2 mitogenic assay (Davis and Linzer 1989b), might also fall into the class of site 2 mutants. However, until now there is no direct evidence that the PRL-R can be dimerized by one ligand molecule. Gertler et al. (1993) could only isolate complexes with a 1:1 molar ratio of prolactin and the extracellular domain of the PRL-R. Clearly, further experiments are required for firmly establishing the prevalent mechanism of dimerization.

2.2.3 Intracellular Signaling of the Receptor

What are the signals produced by the activated receptor? With one exception (Zhang et al. 1990), which was not confirmed in a subsequent publication (Nagano and Kelly 1992), inspection of the sequence of the intracellular domains of the PRL-R did not reveal any sequence motifs that indicate an enzymatic function or a relationship to receptor tyrosine kinases. The same was true for the other receptors of the cytokine superfamily. Very recent work has revealed instead that many, if not all, cytokine receptors have a nonconvalently linked cytoplasmic tyrosine kinase attached to their intracellular domains (Argetsinger et al. 1993). Further work has to establish whether ligand-induced dimerization, followed by activation of attached or intrinsic cytoplasmic tyrosine kinases, is a general principle governing the activation of cytokine receptors and receptor tyrosine kinases. By analogy to the findings with the platelet-derived growth factor (PDGF) receptor tyrosine kinase (Fantl et al. 1992), it is likely that the dimerized PRL-R recruits and activates other signaling proteins in addition to tyrosine kinases. Calcyclin, a cell cycle regulated 10-kDa putative calcium binding protein might be a candidate for such an attached factor (Murphy et al. 1988a). Cross-linking of G proteins to proteins which are immunoprecipitated with antiserum to the PRL-R (Too et al. 1990) suggest that also G-proteins are linked to the PRL-R. The availability of larger quantities of pure PRL-R expressed in insect cells (Cahoreau et al. 1992) should facilitate the identification of proteins with an affinity to the PRL-R by biochemical methods. This will complement functional studies performed with mutated PRL-Rs. An experimental procedure for measuring some of the functional activity of PRL-Rs has already been developed (Leseur et al. 1991; Ali et al. 1992). CHO cells were transiently cotransfected with PRL-R and milk protein gene promoter constructs linked to a chloramphenicol acetyltransferase (CAT) reporter. Measurement of the CAT activity induced by prolactin showed that in this system, the promoter activity was dependend on the form of the cotransfected receptor construct. The long form and the major Nb2 PRL-R form, which is a truncated version of the long receptor form were found to be active, whereas the short form was not functional in this assay.

3 Putative Signaling Pathways Employed by Prolactin for the Regulation of Gene Expression

Effects of the hormone on the tyrosine kinases, protein kinase C (PKC), cyclic nucleotides, intracellular calcium, phospholipases, enzymes regulating the polyamine metabolism and other potential cellular signaling molecules have been investigated extensively. A comprehensive review (Rillema et al. 1988) summarizes earlier work on postreceptor actions. Which of these effects are necessary for the specific expression of prolactin dependent genes and thus play a role in reprogramming the cell after activation of the PRL-R? I will focus here on the evidence for tyrosine kinases, PKC isozymes and intracellular calcium-mediated signaling being involved in regulating pro-lactin-dependent gene expression. Since a complete signal transduction cas-cade intiated by binding of prolactin to its receptor and ultimately effecting the expression of a protein at a transcriptional or posttranscriptional level has not yet been established, our present knowledge about the individual roles of these different signaling systems relies mostly on the results of experiments performed with more or less specific inhibitors.

3.1 Tyrosine Kinases

As discussed in Sect. 2.2.3, one of the first events after binding of the hormone to its receptor is a stimulation of tyrosine kinase activity. The initial observation was made by antiphosphotyrosine immunoblotting experiments. With this technique a rapid phosphorylation of multiple cellular proteins was demonstrated in Nb2 rat lymphoma cells by two independent research groups. A protein designed p120 (Rui et al. 1992) or 121-kDa protein (Rillema et al. 1992) migrated at identical position as a protein induced by growth hormone in 3T3-F442A fibroblasts (Campbell et al. 1993). Sub-sequent work with specific antibodies (Argetsinger et al. 1993) revealed that this protein corresponds to the ubiquitously expressed tyrosine kinase JAK2 (Wilks et al. 1991). This tyrosine kinase belongs to a family which does not contain SH2 or SH3 domains but harbors the structural features of a second protein kinase activity with unknown function. A new molecular weight of 130 kDa was assigned to that protein in this study. It is not known whether JAK2 is the only tyrosine kinase involved in the signaling of the receptor.

The functional role of JAK2 or other tyrosines on activation of gene expression was evaluated with specific inhibitors. Prolactin-dependent milk protein gene expression was abolished by genistein and staurosporin, but not by lavendustin A (Fan and Rillema 1992; Bayat-Sarmadi and Houdebine 1993). Recently, MGF, a mammary gland specific nuclear factor was ident-ified. DNA binding of this factor is induced directly by prolactin (Sect. 6). The activated factor is recognized by antibodies directed against phosphoty-rosine, indicating that it contains phosphotyrosine residues. Genistein and staurosporin but not lavendustin A inhibited the induction (T. Welte, unpub-lished). Thus, at least one of the effects of tyrosine kinases appears to be mediated by MGF. As discussed later, this nuclear factor is related to the STAT family of transcriptional activators, which comprises factors induced by cytokines and peptide growth factors. It is presently not known, which tyrosine kinase is involved in the phosphorylation of MGF, but JAK2 is obviously a candidate.

In addition to MGF and the JAK2 kinase, other proteins have been observed to be phosphorylated on tyrosine in response to prolactin (Rui et al. 1992; Rillema et al. 1992). It is possible that, upon phosphorylation, these other proteins acquire the capability to bind *src* homology domain 2 (SH2 domain) proteins as it has been demonstrated for the autophosphorylated growth factor receptors (Fantl et al. 1992). Some of these SH2 proteins have been shown to be involved in subsequent steps of cellular signaling. In the case of the receptor for epidermal growth factor (EGF-R) or PDGF-R, SH2 domain proteins link phospholipase C-mediated pathways and signaling regulated by p21 ras to the activation of tyrosine kinases. Such mechanisms might be responsible for the effects of prolactin on PKC and on intracellular calcium.

3.2 Protein Kinase C

PKC is known to comprise a large family of isozymes (Nishizuka 1992) with common structural features. There is no unique activation mechanism for the members of the family. The PKC isozymes α, β, and γ can be activated by calcium and phosholipids, isozymes δ, ε, η and ϑ require phospholipids but are calcium insensitive. There is a third group of atypical isozymes which are not activated by phorbol esters and are insensitive to calcium. The isozymes are differentially blocked in their activity by inhibitors and appear to participate in different signaling pathways in a cell type-specific fashion. So far, investigations published about the role of PKC in prolactin signaling did not take into account the heterogeneity of the PKC family.

Activation of PKC after stimulation of cells with prolactin has been observed in various cellular systems: In mammary gland extracts (Waters and Rillema 1989) and hypothalamic slices (DeVito et al. 1991), prolactin triggered within 15 min the redistribution of PKC from the cytosolic to the particulate fraction of cell extracts. In the mouse mammary cell line NOG-8, where prolactin acts as a mitogen, a similar cellular redistribution is observed already after 5 min (Banerjee and Vonderhaar 1992). Activation of a nuclear PKC has been described in rat liver nuclei (Buckley et al. 1988). The effect of inhibitors of PKC on prolactin dependent cellular functions was analyzed in a number of studies. Whereas two reports document an inibitory effect of H7 (Waters and Rillema 1989; Banerjee and Vonderhaar 1992), another study performed with more inhibitors (Bayat-Sarmadi and Houdebine 1993) led to the conclusion that a protein kinase, which is not PKC, is important in mediating the effect of prolactin. In Nb2 cells, H7 was not effective in blocking the early effects of prolactin on stimulation of DNA synthesis (Rillema et al. 1989).

Attempts to mimic the effect of prolactin by a direct stimulation of PKC with the phorbol ester TPA were only partially successful. In Nb2 cells, TPA increased the mitogenic effect induced by the activation of the PRL-R (Gertler et al. 1985). However, in the same cell line, the phosphorylation pattern of strathmin induced by prolactin is distinct from the phosphorylation induced by TPA (Meyer et al. 1992). These results suggest that the protein kinase important for the stimulation of mitogenesis in Nb2 cells was not one of the PKC isoforms which can be activated by TPA. Consistent results indicating a role of PKC in the prolactin signaling of prostate cells were presented (Franklin et al. 1992): Stimulation of mitochondrial aspartate aminotransferase was observed by addition of TPA and prolactin. Both responses were inhibited by gossypol. Downregulation of PKC by treatment of cells with TPA inhibited the induction. In a number of cell lines, long-term treatment with TPA also resulted in a reduction of PRL-R levels (Ormandy et al. 1993).

A final judgment about the role of PKC in the response to prolactin has to await more refined investigations which account for the heterogeneity in the activation mechanisms and the cell type-specific expression pattern of the isozymes.

3.3 Intracellular Calcium

Intracellular calcium ions modulate the activity of several proteins in the cell. Activation of receptors are often found to induce dynamic temporal and spatial changes in the concentration of this intracellular second messenger. Release of calcium from intracellular stores or calcium influx through specific calcium channels in the plasma membrane temporarily increase the concentration of free calcium in the cytosol. Complex oscillation patterns and calcium waves have been observed with sophisticated techniques (Lechleiter and Clapham 1992). Both frequency and amplitude of such oscillations are believed to be important for the mode of cellular response. Present knowledge about the role of such dynamic calcium responses in the signaling of prolactin is very limited. In Nb2 cells, the role of intracellular calcium ions in mediating the action of prolactin has been studied employing calcium ionophores, $^{45}Ca^{2+}$, calcium channel blockers, and intracellular calcium antagonists (Murphy et al. 1988b). The authors conclude that activation of calcium channels was not an important early event in the mitogenic action of prolactin. Instead, effects obtained with the putative intracellular antagonist TMB-8 pointed to a role of an intracellular calcium pool for mediating the mitogenic effect of prolactin in this cell line. Work on the effect of prolactin on hepatocytes of lactating rats (Villalba et al. 1991) led also to the conclusion that mobilization of calcium from intracellular stores is involved in the induction of glycogen phosphorylase-α activity by prolactin. On both cell types, prolactin acts as a mitogen. Prolactin-induced mobilization of calcium from intracellular stores might be restricted to such cells since in the mouse mammary epithelial cell line HC11, where prolactin induces milk protein gene expression in growth-arrested cells (Doppler et al. 1989), no mobilization of intracellular calcium stores was observed with single-cell imaging techniques. However, a rapid prolactin-induced calcium influx from the extracellular space could be demonstrated (T.Bader and K.Maly, unpublished). A blockade of this calcium influx by ethyleneglycol-bis-β-amino-ethylether-N,N,N',N'-tetraacetic acid (EGTA) still allowed the activation of the prolactin-inducible nuclear factor MGF (T.Welte, unpublished), implicating that calcium influx was not required for the activation of this factor. The prolactin-induced increase in intracellular calcium might be important for functions of prolactin other than transcriptional regulation of genes, e.g., regulated exocytosis, which has been shown to be dependent on extracellular calcium in lactating mammary epithelial cells (Turner et al. 1992).

4 Genes Regulated by Prolactin

The number of genes documented to be regulated by prolactin is still small when compared with the great number of physiological actions of the hormone. A summary of such prolactin-induced genes is shown in Table 2. The best studied organ is the mammary gland. For most of the milk proteins (recently reviewed by Rosen et al. 1986, Harris et al. 1990, and Hennighausen et al. 1991) and a number of enzymes involved in the biosynthesis of milk lipids (Barber et al. 1991, 1992a,b) and lactose (Jagoda and Rillema 1991; Peters and Rillema 1992) a positive effect of prolactin on the expression has been demonstrated. This ensures the maintenance of high levels of these proteins during lactation. Lactose synthetase activity requires two separately regulated proteins, galactosyl transferase and α-lactalbumin. Galactosyl transferase is induced by prolactin prior to α-lactalbumin in mouse mammary gland explants (Jagoda and Rillema 1991). Among the various hormones and growth factors secreted into the milk of mammals, EGF has been described to be produced under the control of prolactin (Fenton and Sheffield 1991). Expression of an insulin-like growth factor (IGF) is induced by prolactin and other hormones belonging to the prolactin family (Fielder et al. 1992). By contrast, there is no evidence for such a regulation in bovine mammary tissues (Campbell et al. 1991). The major constituent of the "cropmilk" induced in response to prolactin in pigeons and doves is annexin Icp35 (Hitti and Horseman 1991; Xu and Horseman 1992). The mammalian

Table 2. Prolactin-regulated genes

Tissue	Gene product	Tissue	Gene product
Mammary gland	*Milk proteins* α-Caseins β-Casein WAP β-Lactoglobulin LLP	Liver	PEPCK Synlactin
	Lipid metabolism Acetyl CoA carboxylase Fatty acid synthase Malic enzyme Lipoprotein lipase	Prostate	IGF-I IGF-I receptor Androgen receptor Mitochondrial aspartate aminotransferase (mAAT)
	Lactose synthesis Lactose synthetase Galactosyl transferase α-Lactalbumin Glucose carrier	Nb2 lymphoma	IRF-1 Myc gene product ODC Hsp-70-like protein
	Growth factors and receptors IGF-I-binding protein EGF	Uterus	α₂-Macroglobulin Uteroglobin
	Other proteins Annexin Icp35 120-kDa protein Muc-1 PIP	Ovary	α₂-Macroglobulin Luteinizing hormone receptor

WAP, whey acidic protein; *LLP*, late lactation protein; *IGF-I*, insulin-like growth factor; *EGF*, epidermal growth factor; *PIP*, prolactin-inducible protein; *PEPCK*, phosphoenolpyruvate carboxykinase; *mAAT*, mitochondrial aspartate aminotransferase; *IRF-1*, interferon regulatory factor-1; *ODC*, ornithine decarboxylase.

counterpart of this calcium-dependent, phospholipid-binding protein is also highly expressed in the mammary gland, but is not regulated by prolactin (Horlick et al. 1991). The late-lactation protein (LLP) has been identified as a novel, presumably prolactin-regulated milk protein predominant in the "late milk" of marsupials (Collet et al. 1989, 1991).

In cell lines derived from the mammary gland, three gene products of unknown function have been found to be induced by prolactin. The mammary epithelial cell line HC11 has been shown to accumulate a transcript encoding a 120-kDa protein when treated with prolactin in the presence of the protein synthesis inhibitor cycloheximide (Ball et al. 1988b). This protein has not been characterized further. Muc-1, a highly glycosylated mucin expressed on the surface of mammary epithelial cells has been shown to be induced by prolactin in CID-9 mammary epithelial cells (Parry et al. 1992). Prolactin-inducible protein (PIP) was described as a secreted, prolactin-inducible protein expressed by human breast cancer cells (Shiu and Iwasiow 1985). The cDNA cloning revealed the identity of PIP with the gross cystic disease fluid protein-15 (GCDFP-15) found in human breast cystic fluid (Murphy et al. 1987; Myal et al. 1991).

The mouse mammary tumor virus (MMTV) is expressed at high levels in lactating mammary glands. Enhancer elements in the long terminal repeat (LTR) have been localized which direct the mammary epithelial cell-specific expression (Mink et al. 1992). There is controversy about a positive contribution of prolactin to the mammary-specific expression of viral messenger RNA (mRNA). One report described a stimulation of MMTV-LTR directed expression by prolactin in transiently transfected human breast-carcinoma cells (Haraguchi et al. 1992). However, in transgenic mice (Mok et al. 1992) or stable transfected mouse mammary epithelial cells (Härtig et al. 1993) no evidence for such a regulation was found.

There are fewer reports about prolactin-regulated genes in other organs. In the liver, prolactin increases the mRNA levels of the cytosolic form of phosphoenolpyruvate carboxykinase (PEPCK; Zabala and Garcia-Ruiz 1989) and induces synlactin, a secreted factor that acts synergistically with prolactin on the growth of the pigeon crop sac (Nicoll et al. 1985).

In rat prostate, prolactin enhances the effects of androgens by increasing androgen receptor levels (Prins 1987). In addition, a prolactin-dependent increase of IGF-I message in prostatic stroma together with an increase of IGF-I receptor in the rat prostate was observed, suggesting that prolactin is establishing a paracrine-acting IGF-I system. Citrate production and accumulation have been described to be unique properties of the rat and human prostate. A key regulatory enzyme in this process, the mitochondrial aspartate aminotransferase (mAAT) has been described to be stimulated by prolactin in pig prostate epithelial cells (Franklin et al. 1992).

In Nb2 lymphoma cells prolactin regulates a number of growth related genes (Table 2; de Toledo et al. 1987; Yu-Lee 1990). The interferon regulatory factor IRF-1, a transcription factor which has been found to be inducible by a variety of cytokines (Yu-Lee et al. 1990; Schwarz et al. 1992), belongs to this set of genes and is induced during both early G1 and the G1→S transition (Stevens and Yu-Lee 1992). In the murine T lymphocyte cell line L2 induction of IRF-1 by prolactin at the G1→S transition was also observed. However, the induction of IRF-1 at early G1 was triggered by IL-2, which is necessary for stimulating proliferation in this cell line, and not by prolactin. It is possible that the aberrant expression of PRL-Rs in Nb2 cells (Ali et al. 1991; O'Neal et al. 1991) is responsible for the altered response of this tumor cell line.

In the mesometrial decidua of the uterus (Gu et al. 1992) and in ovary (Gaddykurten and Richards 1991) α2-macroglobulin has been found to be positively regulated by prolactin. This protease inhibitor was previously described as an acute phase protein induced in liver cells by IL-6. Its regulated expression in the decidua might be important for protecting the mesometrium against extensive tissue damage as a result of trophoblast invasion (Gu et al. 1992). The uteroglobin gene has also been found to be induced by prolactin in the uterus (Randall et al. 1991; Kleis-San Francisco et al. 1993). By inducing the levels of the luteinizing hormone receptor in the corpus luteum, prolactin increases the susceptibility of the organ to respond to that hormone (Gåfvels et al. 1992).

There appears to be no unifying mechanism underlying the induction of gene expression by the hormone. Both direct and indirect effects on transcription initiation and posttranscriptional effects have been described. In the regulation of the β-casein gene expression, both mechanisms might be utilized simultaneously (Eisenstein and Rosen 1988; Doppler et al. 1990; Goodman and Rosen 1990). In this review I will focus only on the mechanism by which prolactin stimulates the initiation of transcription, since there is presently no precise information as to how prolactin influences gene expression at other levels of regulation.

5 Search for Prolactin Response Elements

With the availability of genomic clones of milk protein genes, the identification of regulatory *cis*-acting sequences by gene transfer experiments employing truncated and mutated genes and gene fragments appeared to be straightforward. However, due to an initial lack of transfectable cell lines that maintained the capability to produce milk protein genes, the first successful demonstration of the existence of regulatory elements within these genes came from experiments with transgenic mice (reviewed in Hennighausen 1990; Harris et al. 1990). Mammary gland explant cultures of transgenes harboring whey acid protein (WAP) gene promoter constructs revealed that a prolactin response region was contained within a 2.4-kb promoter fragment of the mouse WAP gene (Pittius et al. 1988; Schoenenberger et al. 1990). The first successful demonstration of prolactin response regions by gene transfer experiments employing mammary epithelial cells was done with a 2.3-kb fragment of the rat β-casein gene promoter (Doppler et al. 1989). The novel developed cell line HC11 (Danielson et al. 1984; Ball et al. 1988b) and the selection of stably transfected cells were found to be crucial for an analysis of a prolactin-dependent transcriptional response. With this system, a fine mapping of the prolactin response region to an element extending from −176 to −82 in the rat β-casein gene promoter was possible (Doppler et al. 1989; 1990 and unpublished data). Similar experiments done with the mouse WAP gene promoter defined a minimal hormone response region extending form –165 to +24 (Doppler et al. 1991; Jennewein 1992). Using a similar approach, a prolactin-inducible enhancer region was found between –1676 and –1517 in the bovine β-casein gene (Schmidhauser et al. 1990, 1992). In HC11 cells, attempts to simplify the screening procedure for response elements by employing transient transfection assays failed. Both the β-casein and the WAP gene promoters were expressed at least 10 000 fold weaker in transiently transfected cells, irrespectively of which transfection technique was used (Doppler et al. 1991). However, with primary mouse mammary epithelial cells, it was possible to develop a transient transfection protocol (Yoshimura and Oka 1990a), which served to define prolactin response elements in the mouse β-casein gene between –258 and +7 (Kanai et al. 1993), the rabbit WAP gene between –3000 and +490 (Devinoy et al. 1991) and the rabbit αS1-casein gene between –3768 and –3155 (Pierre et al. 1992). In addition to the systems utilizing mammary epithelial cells, a very elegant technique for the screening of prolactin response elements was published (Lesueur et al. 1990, 1991; Ali et al. 1992). Transient cotransfection of CHO cells with PRL-R expression vectors together with reporter constructs containing potential prolactin response regions were performed. The importance of this type of experiment to define PRL-R regions required for intracellular signaling has already been discussed (Sect. 2.2.3). So far, there have been no reports about the utilization of this system in defining prolactin response regions. A potential caveat in that approach is the employment of a nonmammary epithelial cell for studying the regulation of mammary cell-specific regulated genes. In fact, the first nuclear factor with a documented role in the prolactin response is a mammary cell-specific factor (Sect. 6). Gene transfer of a lactogenic hormone response region into living mice by jet injection of DNA was described recently (Furth et al. 1992). Expression of the injected WAP gene promoter construct was detectable in explant cultures of the mammary glands derived from injected mice. Additional work has to be performed to evaluate the advantages of this novel method in comparison to the established techniques.

6 Mammary Gland Factor MGF:
A Prolactin-Inducible Nuclear Factor

The relatively precise localization of the prolactin response region of the rat β–casein gene promoter allowed a focused analysis of the nuclear factors binding in vitro to that region (Schmitt-Ney et al. 1991). Only one of the factors found was specific for the mammary gland. Because of this property, the factor was termed MGF (mammary gland factor). Mutation of the MGF binding site abolished the function of the β-casein gene promoter in responding to the synergistic action of prolactin and glucocorticoids, implicating the importance of this factor in regulating the inducible gene expression of milk protein genes. Binding sites for this factor were found in both α- and β-casein gene promoters (Table 3). A rat mammary gland preparation with highly purified MGF was size-fractionated by sodium dodecyl sulfate–polyacrylamide gel electrophoresis (SDS-PAGE). MGF binding activity could be assigned to a region of the gel where a single 89-kDa protein was detectable (Wakao et al. 1992). Binding of MGF in vitro was shown to be dependent on phosphorylation. Incubation with phosphatase abolished binding, whereas treatment with casein kinase II was partially able to restore the binding activity (Schmitt-Ney et al. 1992a).

6.1 Mechanism of Activation

Examination of the MGF binding activities at various stages of the mammary gland development was performed (Schmitt-Ney et al. 1992a,b; K. Garimorth, unpublished). Activity was high during lactation and dropped rapidly in the involuting gland. Restimulation of milk synthesis in the glands of weaning mothers by adding back the suckling pups reactivated MGF activity within 90 min. Unexpected, MGF activity was highest at the beginning of lactation and then dropped markedly within the next days (K. Garimorth, unpublished). By contrast, β-casein production is relatively low at the beginning of lactation and continues to increase over a period of several days. When nuclear extracts of HC11 cells stimulated with prolactin and dexamethasone for 4 days were investigated, only very low levels of a factor with slightly slower mobility than MGF but with indistinguishable binding specificity could be detected (Schmitt-Ney et al. 1992a). Later work (T. Welte, unpublished) revealed a rapid down-regulation of MGF binding activity in these cells by prolonged incubation with prolactin. The factor was maximally induced within 15 min after addition of prolactin. Neither glucocorticoid hormone nor insulin was required for the activation process. Cycloheximide did not block the activation, thus de novo protein synthesis was not required. Genistein and staurosporin, two inhibitors of tyrosine kinase activity, inhibited the activation of the factor by prolactin, indicating the involvement of tyrosine kinases. The factor appeared to be a direct substrate of a tyrosine kinase, since phosphotyrosine-specific antibodies and the phosphotyrosine analog phenylphosphate blocked the binding of MGF in vitro (T. Welte, unpublished). The sequences of the published binding sites for MGF are shown in Table 3.

Table 3. Binding sites for MGF, growth hormone-inducible factor, and STAT family members

Binding site	Inducer	Sequence	Reference
Rat β-casein, proximal site	Prolactin	TTCTTGGAATT	Schmitt-Ney et al. 1991
Rat β-casein, distal site		TTCTTGGGAAA	
Bovine β-casein		TTCTAGGAATT	Wakao et al. 1992
Bovine αS1-casein		TTCTTAGAATT	
Consensus of MGF core site:		TTCTYRGRAYY	
Spi 2.1 gene	Growth hormone	TTCTGAGAAAT	Yoon et al. 1990
Rat α2 macroglobulin (APRE)	IL-6	TTCTGGGAATT	Wegenka et al. 1993
FcγRI (GRR)	INF-γ	TTCTGGGAAAT	Pearse et al. 1993
Ly-6E (GAS)	INF-γ	TTACAGGAATA	Khan et al. 1993
Human *c-fos* (SIE)	EGF, PDGF	TTACGGGAAAT	Hayes et al. 1987

MGF, mammary gland factor; *IL*, interleukin; *INF*, interferon; *EGF*, epidermal growth factor; *PDGF*, platelet-derived growth factor.

6.2 Relationship to Other Inducible Nuclear Factors

The sequence motif of MGF matches the motifs published for a number of growth factor and cytokine inducible proteins. A growth hormone-inducible DNA-binding protein has been identified in rat liver (Yoon et al. 1990). It binds to a growth hormone response region in the serine protease inhibitor (Spi) 2.1 gene. The sequence of the binding region contains a motif highly related to the MGF consensus (Table 3). Induction of the factor was per-fomed by injection of human growth hormone into hypophysectomized and hypothyroid male rats. As already mentioned in Sect. 2, human growth hormone has both somatogenic and lactogenic activity. Thus further experi-ments are required to rule out the possiblity that an activation of PRL-Rs by human growth hormone is responsible for the induction. Very similar bind-ing sites have also been described for the EGF and *sis*-inducible factor (SIF) (Hayes et al. 1987), the interferon-γ inducible transcription factor complexes (Pearse et al. 1993; Wilson and Finbloom 1992; Khan et al. 1993), and the IL-6-inducible acute phase response factor APRF (Wegenka et al. 1993).

Common principles in the mechanism of the activation of SIF, APRF, the factors induced by interferon-α, -β and -γ and other cytokines have been described recently (Schindler et al. 1992; Larner et al. 1993; Ruff-Jamison et al. 1993; Silvennoinen et al. 1993; Sadowski et al. 1993). The well-studied mechanism by which the signal transducer and activator of transcription Stat91 induces transcription serves as a paradigm. This 91-kDa protein belonging to the ISGF3 γ family contains a SH2 domain. The protein is rapidly phosphorylated on tyrosine 701 in response to interferon-γ. This phosphorylation is necessary for the activation of the factor (Shuai et al. 1993). Only the phosphorylated protein is bound to the γ-interferon response region (GRR) as part of the transcription factor complex. Whether Stat91 alone is sufficient for binding or whether additional peptides are required is presently unknown. Stat91 is also part of some, but not all, of the other transcription factor complexes induced by cytokines and peptide growth factors investigated so far. The site of phosphorylation is presumable a membrane-bound complex of the factor and the activated receptor (David et al. 1993).

MGF appears to belong to the same family of factors. It binds to the same or a closely related DNA recognition sequence and has a similar mode of activation, namely, phosphorylation on phosphotyrosine. Electromobility shift experiments indicate that the mobility of the MGF complex is different from the interferon-γ inducible transcription factor complex but similar to the IL-6 induced factor APRF (T. Welte , unpublished). The molecular cloning of the component(s) of the factors should give further insights into the relationship of MGF to cytokine-induced factors in the near future.

7 Synergy and Interference of Other Signaling Pathways with Prolactin Action

A frequent finding characteristic for prolactin-induced gene expression is the dependence on other activated signaling pathways (Table 4). This appears to be one important control mechanism to restrict a specific prolactin response to appropriate tissues and stages of development.

7.1 Steroid Hormones

Synergistic and antagonistic interactions of prolactin signaling pathways with steroid hormone-dependent cellular responses are frequently observed, especially in the mammary gland (reviewed by Topper and Freeman 1980). Glucocorticoid hormones have in most cases a pronounced synergistic effect on prolactin-dependent milk protein gene expression. The degree of synergism is not the same for all milk protein genes (Ono and Oka 1980; Ray et al. 1986; Puissant and Houdebine 1991). This might contribute to the noncoordinate regulation of milk proteins in the mammary gland. The synergism between glucocorticoid hormones and prolactin also can be observed in the regulation of stable transfected milk protein gene promoter constructs (Doppler et al. 1989, 1991, Schmidhauser et al. 1992), indicating that the two hormones both act on the level of transcription. Again, the mode of synergism was not the same for different milk protein gene regulatory elements. In the case

Table 4. Factors influencing the effect of prolactin on gene expression

	Factor	Effect
Steroids	Glucocorticoids	+
	Progestins	– or +
	Estrogens	+
	Androgens	+
	Thyroxine	+
Peptide factors	Insulin/IGF-1	+
	Growth hormone	+
	EGF/TGF-α	– or +
	TGF-β	–
	IL-2	+
Cellular organization	Cell–cell interactions	+
	Interactions with basement membrane constituents	+
	Establishment of the cellular polarity	+
	Accessability of hormonal receptors	+

IGF, insulin growth factor; *EGF,* epidermal growth factor; *TGF,* transforming growth factor; *IL,* interleukin.

of the rat β-casein gene promoter only in the presence of both hormones an effective activation of transcription of the linked reporter was possible. By contrast, the WAP gene promoter was induced considerably by glucocorticoids alone (Doppler et al. 1991) and the lactogenic hormone enhancer of the bovine β-casein was only weakly dependent on the steroid hormone. In the WAP gene promoter, the glucocorticoid effect alone was found to be variable and appeared to depend on the integration site of the transfected promoter (Jennewein 1992). Mammary gland explant cultures of transgenic mice which harbored WAP promoter constructs juxtaposed to matrix-attachment regions showed a strict dependence on both glucocorticoid hormone and prolactin for induction of the transgene (McKnight et al. 1992).

What is the mechanism of synergy between glucocorticoid hormones and prolactin? Analysis of the kinetics of hormonal induction of the endogenous β-casein gene and a stable, transfected β-casein gene promoter construct revealed a fast response to prolactin and a slow response to glucocorticoid hormone (Doppler et al. 1991). The delayed effect of glucocorticoid hormones and the absence of classical glucocorticoid binding sites in the hormone response region of the β-casein gene promoter pointed to an indirect effect, requiring de novo synthesis of proteins. Indeed, such requirement has been demonstrated for the mouse β-casein gene expression (Yoshimura and Oka 1990b). Investigations performed with purified rat liver glucocorticoid receptor preparations (Welte et al. 1993) have identified nonclassical glucocorticoid receptor binding sites in the hormone response regions of the WAP gene promoter and the β-casein gene promoter. These sites are related to the delayed secondary glucocorticoid receptor binding sites found in the regulatory region of the α2u-globulin gene (Chan et al. 1991) and might also be involved in the delayed response of the milk protein genes to glucocorticoids.

Progesterone has been found to act negatively or positively on the regulation of prolactin-dependent genes, depending on the tissue and on the developmental stage. In the mammary gland the hormone is described as an inhibitor of prolactin-induced milk protein synthesis at the end of pregnancy, when progesterone levels are high (Sankaran and Topper 1988; Jahn et al. 1989). A nuclear factor induced by progesterone which binds to the mouse β-casein gene promoter has been identified (Lee and Oka 1992a,b). Mutational analysis revealed a functional role of this factor in repressing the lactogenic hormone response of the promoter. It is possible that in addition to these factors, the activated progesterone receptor acts directly on the β-casein gene promoter. The recently identified glucocorticoid receptor binding sites (Welte et al. 1993) might also be utilized by the progesterone receptor, thereby antagonizing the action of the glucocorticoid receptor. In contrast to the negative effect of progesterone on the terminal differentiation of the gland, in some cases progestins are believed to increase the proliferative effects of prolactin on mouse mammary epithelium cells, human tumor cell lines, and pregnancy-dependent mouse mammary tumors (Vonderhaar and Biswas 1987; Sakai et al. 1990). A servomechanism of prolactin and progesterone has been proposed to regulate uteroglobin expression in the uterus of rabbits (Chilton et al. 1988). The positive effects of prolactin on uteroglobin induction is thereby amplified by a progesterone-dependent increase of the PRL-R levels.

Numerous effects of other steroid hormones on prolactin action have been described; many of them appear to occur at the level of modulation of receptor expression (Ormandy and Sutherland 1993). Androgens act synergistically with prolactin in the human tumor cell line T47-D on the induction of PIP (Murphy et al. 1987) and increase PRL-R binding sites in a number of tumor cell lines (Ormandy et al. 1992). Estrogens are potent inducers of the PRL-R in the liver of female rats (Posner et al. 1974). This early observation greatly facilitated the cloning of the receptor (Boutin et al. 1988). α-Lactalbumin but not α-casein synthesis was enhanced by treatment of hypophysectomized pregnant mice with a combination of corticosterone and thyroxine (Thordarson et al. 1992), indicating that the expression of α-lactalbumin has a selective requirement for thyroid hormone.

7.2 Peptide Factors

The synthesis of milk protein genes is dependent on insulin (Topper and Freeman 1980). IGF-I can be substituted for insulin in inducing milk protein synthesis (Prosser et al. 1987). Whether the effect is mediated by the insulin receptor or the receptor for IGF-I is unknown. The mode of synergy between prolactin and insulin is presently unclear. Insulin was not required for the transcriptional effects of prolactin on the expression of stably transfected milk protein gene promoter constructs (Doppler et al. 1989) or for the activation of MGF (T. Welte , unpublished).

EGF has been shown to inhibit lactogenic hormone-dependent milk protein gene synthesis and transcription (Taketani and Oka 1983; Hynes et al. 1990; Doppler et al. 1991). Expression of transforming growth factor α (TGF-α), which binds to the same receptor as EGF, was also blocking the transcriptional induction of the milk protein β-casein (Hynes et al. 1990). A blockade of MGF induction by long-term exposure of cells to prolactin and EGF was reported in HC11 cells (Schmitt-Ney et al. 1992b). However, a short-term exposure of the same cells to EGF did not inhibit the transient activation of MGF by prolactin (T. Welte, unpublished), indicating that the inhibitory action of EGF was not a result of a direct interference with a prolactin-induced signaling pathway. Cultivation of HC11 cells in EGF had a positive effect on lactogenic hormone action, but only when the growth factor was present in the medium prior to the addition of lactogenic hormones and was removed concomitantly with the addition of homones (Taverna et al. 1991).

TGF-β blocked the lactogenic hormone induced synthesis of β-casein (Mieth et al. 1990; Robinson et al. 1993). The effect was posttranscriptional and did not affect the level of casein mRNA in explant cultures. Since expression of TGF-β2 and TGF-3 was found to be high during pregnancy but reduced at the onset of lactation, it was hypothesized that these factors are physiological inhibitors of an accumulation of milk proteins prior to term (Robinson et al. 1993).

In the T-cell line L2, IL-2 has been found to be a comitogen, together with prolactin (Clevenger et al. 1990). Analysis of the expression pattern of cell cycle-related proteins after stimulation with IL-2, prolactin, or both hormones led to the preposition of a sequential model of T-cell activation, requiring IL-2 for the early G1$\rightarrow$G1 transition and prolactin, as a progression factor, for a late G1$\rightarrow$S phase transition (Clevenger et al. 1992).

7.3 Cellular Organization

Many of the cell types which have an intrinsic capability to respond express prolactin responsive genes only under special culture conditions. For mammary epithelial cells, the culture conditions which are required for an efficient synthesis and secretion of milk protein genes have been extensively characterized (Strange et al. 1991 and references therein). Cell–cell interactions promoted by cocultivation of nonepithelial with epithelial cell lines, cell interactions with basement membrane constituents and/or establishment of the cellular polarity and the accessibility of hormone receptors have been found to promote hormone-dependent milk protein synthesis. Effects on transcription (Schmidhauser et al. 1992), posttranscriptionally regulated message accumulation (Eisenstein and Rosen 1988; Zeigler and Wicha 1992), and secretion (Strange et al. 1991) of milk protein genes were observed. The epithelial cell lines 31E (Reichmann et al. 1989, Strange et al. 1991) and HC11 (Ball et al. 1988b) did not dependent on exogenously added basement membrane components, indicating that they are able to synthesize these components by themselves. Cultivation of the 31E cells in a filter chamber allowed the formation of electrically tight polarized epithelial cell layers. They were only responsive to prolactin added to the basal chamber, indicating a restriction of PRL-R expression to the basal cell surface. The cell exhibited a polarized secretion of laminin into the lower chamber and of β-casein into the apical chamber (Strange et al. 1991).

8 Dysregulation of Prolactin-Dependent Gene Expression

Until recently, very little was known about the physiological mechanisms by which prolactin activates gene epression. Thus, only a limited evaluation of aberrant regulatory mechanisms was possible so far.

8.1 Effects of Oncogene Expression

The availability of prolactin-responsive clonal cell lines allowed the assessment of effects induced by the expression of transfected or infected oncogenes. The only reported positive effect on the response to lactogenic hormones was observed by introducing *v-myc* into a mouse mammary epithelial cell line (Ball et al. 1988a). Nuclear extracts of tumor cells derived from WAP-*myc* transgenes exhibited high constitutive levels of a factor which migrated at a different position in comparison to MGF found in pregnant mice (Happ et al. 1993). Whether this factor exhibits the same binding specificties as MGF was not shown. Experiments with *v-myc* infected HC11 cells did not show an effect on lactogenic hormone dependent β-casein gene promoter activity (Jehn et al. 1992). Thus, further experiments are required to establish the role of *myc* oncogene products in the cellular response to prolactin.

Transfection or infection with an activated *Ha-ras* oncogene consistently decreased the response to lactogenic hormones (Hynes et al. 1990; Jehn et al. 1992; Happ et al. 1993). Evidence for two different sites of the inhibitory action has been presented. In *Ha-ras* transfected HC11 cells, no MGF was detectable after a 2-day induction with lactogenic hormones (Happ et al. 1993), implicating an effect of activated p21 ras on the prolactin signaling. Cellular synthesis of p79 gag-raf, an oncogene product implicated to act downstream of p21ras, had similar effects. However, before definite conclusions can be drawn, experiments have to be done which compare the effects of *Ha-ras* and *v-raf* on the strong transient activation of MGF by prolactin. Another report describes a negative effect of *Ha-ras* on the glucocorticoid receptor-mediated signaling by increasing the levels of the transcription factor AP-1 (Jehn et al. 1992). A similar mechanism was proposed by the same authors to be responsible for the impaired lactogenic hormone response of *v-src* and *mos* expressing HC11 cells. It is possible that the inhibitory effects of EGF are also mediated by p21 ras, since stimulation of the EGF receptor results in an increase of the active p21 ras-GTP (Grunicke and Maly 1993). An additional potential connection between p21 ras and EGF action are the observed elevated TGF-α levels in the culture medium of *Ha-ras* transfected HC11 cells (Hynes et al. 1990), which suggests the establishment of an autocrine loop by the activated oncogene.

Expression of the oncogenes *int-2*, activated rat *neu*, or T-antigen in HC11 cells did not affect the response to prolactin (Hynes et al. 1990; Wolff et al. 1992; Happ et al. 1993). However, all transfectants acquired the phenotype of transformed cells, indicating that transformation of cells per se was not sufficient to change the response to lactogenic hormones.

8.2 Aberrant Prolactin Signaling in Tumors

The role of dysregulated prolactin-dependent pathways in tumor formation was addressed mainly by studying mammary gland tumors. Measurement of prolactin hormone levels in the serum of tumor patients (Di Carlo et al. 1988; Maddox et al. 1992; Adams 1992), and the determination of PRL-R binding sites in tumors and tumor cell lines (Shiu 1979; Bonneterre et al. 1987; DiCarlo et al. 1988; Holtkamp et al. 1988; Bonneterre and Peyrat 1989; De Placido et al. 1990) were performed. The studies revealed no conclusive evidence for aberrant prolactin signaling in human tumors. In addition, pharmacological blockade of prolactin secretion (Kleinberg 1987) was not effective in repressing the growth of human tumors. The novel insights in the regulation of prolactin gene expression should stimulate a reevaluation. Instead of the insensitive screening for binding sites in tumors with radioactively labeled hormone, the availability of the cloned receptor will allow a measurement of receptor mRNA abundancy and structure and will facilitate the development of receptor-specific antibodies. It is possible that these studies will reveal the existence of aberrant human receptor forms. In the rat tumors, two cases with aberrant receptor expression already have been documented. First, Nb2 lymphoma cells have a truncated version of the long form of the receptor (Ali et al. 1991). Second, an increased expression of the PRL-R without recognizable change in the structure of the protein was detected in a Moloney murine leukemia virus induced rat thymoma. The activation of the receptor gene was the result of the insertion of the 5' long terminal repeat in front of the PRL-R (Barker et al. 1992). Thus in T-cells, the PRL-R may function as an oncogene. The development of prolactin receptor-specific antagonists (Davis and Linzer 1989b; Fuh et al. 1993) should facilitate a selective blockade of the PRL-R. The pharmacological blockade of prolactin secretion done so far is only partially effective in this respect, since it does not inhibit the synthesis of human growth hormone (Köbberling 1983), which can also bind and activate the PRL-R.

An investigation of alterations of prolactin-dependent gene expression in human tumors has not been conducted yet. The analysis of the protein composition of cyst fluids revealed a change of the expression pattern of milk proteins in samples of breast tumor patients (Vizoso et al. 1992). Whether this change was due to an altered response to prolactin is unknown. GCDEFP-15 and Muc-1, two prolactin-inducible proteins (Sect. 4), are frequently found in tumors and have already been used as tumor markers before their regulation by prolactin was discovered. Again, the potential role of prolactin in causing the high levels of these proteins in some tumors is unclear. In the future, a better understanding of the intracellular signaling employed by prolactin and the cloning of the gene or the genes encoding the prolactin responsive nuclear factor MGF will facilitate the detection and understanding of a dysregulated expression of prolactin-dependent genes in tumor cells.

9 Summary and Conclusions

The molecular characterization of the receptor for prolactin and the identification of a nuclear factor involved in the transcriptional activation of prolactin dependent genes have greatly increased our understanding about the mechanism by which this multifunctional hormone prolactin activates gene expression. At least in the activation of milk protein gene expression, the following sequence of events appears to be initiated by the binding of prolactin to its receptor: (1) ligand-induced receptor dimerization; (2) activation of one or more cytoplasmic tyrosine kinases attached to the receptor; (3) phosphorylation of the prolactin-inducible nuclear factor MGF on tyrosine; (4) binding of MGF to its response element on the DNA. This mode of action is very similar to the mechanisms employed by cytokine and growth factor receptors. In addition to the similarities in the mode of action, the receptors and nuclear factors are also structurally related. There is significant sequence homology between the PRL-R, the GH-R, the majority of cytokine receptors, and the interferon receptors. Also the DNA recognition motifs utilized by MGF and nuclear factors induced by interferons, cytokines, and peptide growth factors are highly conserved. This points to an evolutionary relationship of the signaling pathways employed by peptide growth factors, prolactin, growth hormone, a number of cytokines, and the interferons.

The multifunctional activity of prolactin has its biochemical correlate in the developmental and stage-specific control of the receptor and receptor subtypes and in the differential dependence on the activation of other, mostly steroid hormone-controlled signaling systems. It is also likely that there are cell type-specific differences in the signaling pathways coupled to the PRL-R. An additional way to control different physiological actions by the same receptor signaling system is achieved by the temporal and spatial regulation of expression of multiple genes encoding proteins which bind with high affinity to the PRL-R. With the PRL-R and the prolactin-inducible nuclear factor MGF two proteins involved at the beginning and at the endpoint of prolactin-dependent gene regulation have been identified. The further molecular characterization of these proteins will facilitate a biochemical approach to identify other cellular regulators important for the action of prolactin.

Acknowledgements. I thank Drs. K. Maly, T. Welte, P. Jennewein, and J. Lechner for the critical reading of the manuscript. Part of the work done in the laboratory of the author was supported by the Fonds zur Förderung der wissenschaftlichen Forschung, projects P9346 and F209.

References

Adams JB (1992) Human breast cancer: concerted role of diet, prolactin and adrenal C_{19}-Δ^5-steroids in tumorgenesis. Int J Cancer 50:854–858

Ali S, Pellegrini I, Kelly PA (1991) A prolactin-dependent immune cell line (Nb2) expresses a mutant form of prolactin receptor. J Biol Chem 266:20110–20117

Ali S, Edery M, Pellegrini I, Lesueur L, Paly J, Djiane J, Kelly PA (1992) The Nb2 form of prolactin receptor is able to activate a milk protein gene promoter. Mol Endocrinol 6:1242–1248

Amit T, Hochberg Z, Waters MJ, Barkey RJ (1992) Growth hormone-binding and prolactin-binding proteins in mammalian serum. Endocrinology 131:1793–1803

Arey BJ, Averill RLW, Freeman ME (1989) A sex-specfic endogenous stimulatory rhythm regulating prolactin secretion. Endocrinology 124:119–123

Argetsinger LS, Campell GS, Yang X, Witthuhn BA, Silvennoinen O, Ihle JN, Carter-Su C (1993) Identification of JAK2 as a growth hormone receptor-associated tyrosine kinase. Cell 74:237–244

Ball RK, Ziemiecki A, Schoenenberger CA, Reichmann E, Redmond SMS, Groner B (1988a) v-*myc* alters the response of a cloned mouse mammary epithelial cell line to lactogenic hormones. Mol Endocrinol 2:133–142

Ball RK, Friis RR, Schoenenberger CA, Groner B (1988b) Prolactin regulation of β-casein gene expression and of a cytosolic 120-kd protein in a cloned mouse mammary epithelial cell line. EMBO J 7:2089–2095

Banerjee R, Vonderhaar BK (1992) Prolactin-induced protein kinase-C activity in a mouse mammary cell line (NOG-8). Mol Cell Endocrinol 90:61–67

Barber MC, Finley E, Vernon RG (1991) Mechanisms whereby prolactin modulates lipogenesis in sheep mammary gland. Horm Metab Res 23:143–145

Barber MC, Clegg RA, Finley E, Vernon RG, Flint DJ (1992a) The role of growth hormone, prolactin and insulin-like growth factors in the regulation of rat mammary gland and adipose tissue metabolism during lactation. J Endocrinol 135:195–202

Barber MC, Travers MT, Finley E, Flint DJ, Vernon RG (1992b) Growth-hormone prolactin interactions in the regulation of mammary and adipose-tissue acetyl-coA carboxylase activity and gene expression in lactating rats. Biochem J 285:469–475

Barker CS, Bear SE, Keler T, Copeland NG, Gilbert DJ, Jenkins NA, Yeung RS, Tsichlis PN (1992) Activation of the prolactin receptor gene by promoter insertion in a Moloney murine leukemia virus-induced rat thymoma. J Virol 66:6763–6768

Bayat-Sarmadi M, Houdebine L-M (1993) Effect of various protein kinase inhibitors on the induction of milk protein gene expression by prolactin. Mol Cell Endocrinol 92:127–134

Bazan JF (1989) A novel family of growth factor receptors: a common binding domain in the growth hormone, prolactin, erythropoetin and IL-6 receptors, and the p75 IL-2 receptor β-chain. Biochem Biophys Res Commun 164:788–795

Bazan JF (1990) Structural design and molecular evolution of a cytokine receptor superfamily. Proc Natl Acad Sci USA 87:6934–6938

Bchini O, Andres AC, Schubaur B, Mehtali M, LeMeur M, Lathe R, Gerlinger P (1991) Precocious mammary gland development and milk protein synthesis in transgenic mice ubiquitously expressing human growth hormone. Endocrinology 128:539–546

Bonneterre J, Peyrat JP, Beuscart R, Lefebvre J, Demaille A (1987) Prognostic significance of prolactin receptors in human breast cancer. Cancer Res 47:4724–4728

Boutin J-M, Jolicoeur C, Okamura H, Gagnon J, Edery M, Shirota M, Banville D, Dusanterfourt I, Djiane J, Kelly PA (1988) Cloning and expression of the rat prolactin receptor, a member of the growth hormone/prolactin receptor gene family. Cell 53:69–77

Boutin J-M, Edery M, Shirota M, Jolicoeur C, Lesueur L, Ali S, Gould D, Dijane J, Kelly PA (1989) Identification of a cDNA encoding a long form of prolactin receptor in human hepatoma and breast cancer cells. Mol Endorinol 3:1455–1461

Brelje TC, Scharp DW, Lacy PE, Orgen L, Talamantes F, Robertson M, Friesen HG, Sorenson RL (1993) Effect of homologous placental lactogens, prolactins, and growth hormones on islet B-cell division and insulin secretion in rat, mouse, and human islets: implication for placental lactogen regulation of islet function during pregnancy. Endocrinology 132:879–887

Buck K, Vanek M, Groner B, Ball RK (1992) Multiple forms of prolactin receptor messenger ribonucleic acid are specifically expressed and regulated in murine tissues and the mammary cell line-HC11. Endocrinology130:1108–1114

Buckley AR, Crowe PD, Russel DH (1988) Rapid activation of protein kinase C in isolated rat liver nuclei by prolactin, a known hepatic mitogen. Proc Natl Acad Sci USA 85:8649–8653

Cahoreau C, Petridou B, Cerutti M, Djiane J, Devauchelle G (1992) Expression of the full-length rabbit prolactin receptor and its specific domains in baculovirus infected insect cells. Biochimie 74:1053–1065

Campbell GS, Christian LJ, Carter-Su C (1993) Evidence for involvement of the growth hormone receptor-associated tyrosine kinase in actions of growth hormone. J Biol Chem 268:7427–7434

Campbell PG, Skaar TC, Vega JR, Baumrucker CR (1991) Secretion of insulin-like growth factor-I (IGF-I) and IGF-binding proteins from bovine mammary tissue in vitro. J Endocrinol 128:219–228

Chan GC-K, Hess P, Meenakshi T, Carlstedt-Duke J, Gustafsson J-Å, Payvar F (1991) Delayed secondary glucocorticoid response elements: unusual nucleotide motifs specify glucocorticoid receptor binding to transcribed regions of α_{2u}-globulin DNA. J Biol Chem 266:22634–22644

Chilton BS, Mani SK, Bullock DW (1988) Servomechanism of prolactin and progesterone in regulating uterine gene expression. Mol Endocrinol 2:1169–1175

Chiu S, Koos RD, Wise PM (1992) Detection of prolactin receptor (PRL-R) messenger RNA in the rat hypothalamus and pituitary gland. Endocrinology130: 1747–1749

Clapp C, Weiner RI (1992) A specific, high affinity, saturable binding site for the 16-kilodalton fragment of prolactin on capillary endothelial cells. Endocrinology 130:1380–1386

Clarke DL, Linzer DIH (1993) Changes in prolactin receptor expression during pregnancy in the mouse ovary. Endocrinology 133:224–232

Clevenger CV, Russell DH, Appasay PM, Prystowky MB (1990) Regulation of interleukin 2-driven T-lymphocyte proliferation by prolactin. Proc Natl Acad Sci USA 87:6460–6464

Clevenger CV, Altmann SW, Prystowsky MB (1991) Requirement of nuclear prolactin for interleukin-2-stimulated proliferation of T lymphocytes. Science 253:77–79

Clevenger CV, Sillman AL, Hanleyhyde J, Prystowsky MB (1992) Requirement for prolactin during cell cycle regulated gene expression in cloned T-lymphocytes. Endocrinology 130:3216–3222

Collet C, Joseph R, Nicholas K (1989) Molecular cloning and characterization of a novel marsupial milk protein gene. Biochem Biophys Res Commun 164:1380–1383

Collet C, Joseph R, Nicholas K (1991) A marsupial beta-lactoglobulin gene-characterization and prolactin-dependent expression. J Molecular Endocrinol 6:9–16

Crumeyrolle-Arias M, Latouche J, Jammes H, Djiane J, Kelly PA, Reymond MJ, Haour F (1993) Prolactin receptors in the rat hypothalamus: autoradiographic localization and characterization. Neuroendocrinology 57:457–466

Cunningham BC, Mulkerrin MG, Wells JA (1990) Zinc mediation of the binding of human growth hormone to the human prolactin receptor. Science 250: 1709–1712

Danielson KG, Oborn CJ, Durban EM, Butel EM, Butel JS, Medina D (1984) Epithelial mouse mammary cell line exhibiting normal morphogenesis in vivo and functional differentiation in vitro. Proc Natl Acad Sci USA 81:3756–3760

Dardenne M, Kelly PA, Bach JF, Savino W (1991) Identification and functional activity of prolactin receptors in thymic epithelial cells. Proc Natl Acad Sci USA 88:9700–9704

David M, Romero G, Zhang Z-Y, Dixon JE, Larner AC (1993) In vitro activation of the transcription factor ISGF3 by interferon α involves a membrane-associated tyrosine phosphatase and tyrosine kinase. J Biol Chem 268:6593–6599

Davis JA, Linzer DIH (1988) Autocrine stimulation of Nb2 cell proliferation by secreted, but not intracellular, prolactin. Mol Endocrinol 2:740–746

Davis JA, Linzer DIH (1989a) Expression of muliple forms of the prolactin receptor in mouse liver. Mol Endocrinol 3:674–680

Davis JA, Linzer DIH (1989b) A mutant lactogenic hormone binds, but does not activate, the prolactin receptor. Mol Endocrinol 3:949–953

Demeyts P (1992) Structure of growth hormone and its receptor – an unexpected stoichiometry. Trends Biochem Sci 17:169–170

De Placido S, Gallo C, Perrone F, Marinelli A, Pagliarulo C, Carlomagno C, Petrella G, Distria M, Delrio G, Bianco AR (1990) Prolactin receptor does not correlate with oestrogen and progesterone receptors in primary breast cancer and lacks prognostic significance – 10 year results of the Naples Adjuvant (Gun) Study. Br J Cancer 62:643–646

de Toledo, SM, Murphy, LJ, Hatton TH, Friesen HG (1987) Regulation of 70-kilodalton heat-shock-like messenger ribonuleic acid in vitro and in vivo by prolactin. Mol Endocrinol 1:430–434

Devinoy E, Maliénou-N'Gassa R, Thépot D, Puissant C, Houdebine LM (1991) Hormone responsive elements within the upstream sequences of the rabbit whey acidic protein (WAP) gene direct chloramphenicol acetyl transferase (CAT) reporter gene expression in transfected rabbit mammary cells. Mol Cell Endocrinol 81:185–193

DeVito WJ, Stone S, Avakian C (1991) Prolactin stimulation of protein kinase C activity in the rat hypothalamus. Biochem Biophys Res Commun 176:660–667

De Vos AM, Ultsch M, Kossiakoff AA (1992) Human growth hormone and extracellular domain of its receptor: crystal structure of the complex. Science 255: 306–312

Di Carlo R, Muccioli G, Bellussi G, Conti G, Racca S (1988) High tumour prolactin receptor content and lack of increase in serum prolactin levels as predictors of good response to endocrine therapy in rat mammary cancer. Int J Cancer 41:767–770

Di Carlo R, Muccioli G, Papotti M, Bussolati G (1992) Characterization of prolactin receptor in human brain and choroid plexus. Brain Res 570:341–346

Di Mattia GE, Gellersen B, Duckworth ML, Friesen HG (1990) Human prolactin gene expression: the use of an alternative non-coding exon in decidua and in the IM9-P3 lymphoblast cell line J Biol Chem 256:16412–16421

Dombrowicz D, Sente B, Closset J, Hennen G (1992) Dose-dependent effects of human prolactin on the immature hypophysectomized rat testis. Endocrinology 130:695–700

Doppler W, Groner B, Ball RK (1989) Prolactin and glucocorticoid hormones synergistically induce expression of transfected rat β-casein gene promoter constructs in a mammary epithelial cell line. Proc Natl Acad Sci USA 86:104–108

Doppler W, Höck W, Hofer P, Groner B, Ball RK (1990) Prolactin and glucocorticoid hormones control transcription of the β-casein gene by kinetically distinct mechanisms. Mol Endocrinol 4:912–919

Doppler W, Villunger A, Jennewein P, Brduscha K, Groner B, Ball RK (1991) Lactogenic hormone and cell type-specific control of the whey acidic protein gene promoter in transfected mouse cells. Mol Endocrinol 5:1624–1632

Dusanter-Fourt I, Gaye P, Belair L, Petridou B, Kelly PA, Djiane J (1991) Prolactin receptor gene expression in the rabbit – identification, characterization and tissue distribution of several prolactin receptor messenger RNAs encoding a unique precursor. Mol Cell Endocrinol 77:181–192

Edery M, Jolicoeur C, Levi-Meyrueis C, Dusanter-Fourt I, Pétridou B, Boutin J-M, Lesueur L, Kelly PA, Dijane J (1989) Identification and sequence analysis of a second form of prolactin receptor by molecular cloning of complementary DNA from rabbit mammary gland. Proc Natl Acad Sci USA

Eisenstein RS, Rosen JM (1988) Both cell substratum and hormonal regulation of milk protein gene expression are exerted primarily at the posttranscriptional level. Mol Cell Biol 8:3183–3190

Elberg G, Kelly PA, Djiane J, Binder L, Gertler A (1990) Mitogenic and binding properties of monoclonal antibodies to the prolactin receptor in Nb_2 rat lymphoma cells: selective enhancement by anti-mouse IgG. J Biol Chem 265:14770–14776

Fan GA, Rillema JA (1992) Effect of a tyrosine kinase inhibitor, genistein, on the actions of prolactin in cultured mouse mammary tissues. Mol Cell Endocrinol 83:51–55

Fantl WJ, Escobedo JA, Martin GA, Turck CW, del Rosario M, McCormick F, Williams LT (1992) Distinct phosphotyrosines on a growth factor receptor bind to specific molecules that mediate different signaling pathways. Cell 69:413–423

Fenton SE, Sheffield LG (1991) Lactogenic hormones increase epidermal growth factor messenger RNA content of mouse mammary glands. Biochem Biophys Res Commun 181:1063–1069

Ferrara N, Clapp C, Weiner R (1991) The 16 K fragment of prolactin specifically inhibits basal or fibroblast growth factor stimulated growth of capillary endothelial cells. Endocrinology 129:896–900

Fielder PJ, Thordarson G, English A, Rosenfeld RG, Talamantes F (1992) Expression of a lactogen-dependent insulin-like growth factor-binding protein in cultured mouse mammary epithelial cells. Endocrinology 131:261–267

Franklin RB, Ekiko DB, Costello LC (1992) Prolactin stimulates transcription of aspartate aminotransferase in prostate cells. Mol Cell Endocrinol 90:27–32

Freemark M, Comer M (1989) Purification of a distinct placental lactogen receptor, a new member of the growth hormone/prolactin receptor family. J Clin Invest 83:883–889

Fuh G, Colosi P, Wood WI, Wells JA (1993) Mechanism-based design of prolactin receptor antagonists. J Biol Chem 268:5376–5381

Furth PA, Shamay A, Wall RJ, Henninghausen L (1992) Gene transfer into somatic tissues by jet injection. Anal Biochem 205:365–368

Gaddykurten D, Richards JS (1991) Regulation of α2-macroglobulin by luteinizing hormone and prolactin during cell differentiation in the rat ovary. Mol Endocrinol 5:1280–1291

Gåfvels M, Bjurulf E, Selstam G (1992) Prolactin stimulates the expression of luteinizing hormone/chorionic gonadotropin receptor messenger ribonucleic acid in the rat corpus luteum and rescues early pregnancy from bromocriptin-induced abortion. Biol Reprod 47:534–540

Gala RR (1991) Prolactin and growth hormone in the regulation of the immune system. Proc Soc Exp Biol Med 198:513–527

Gellerson B, Bonhoff A, Hunt N, Bohnet HG (1991) Decidual-type prolactin expression by the human myometrium. Endocrinology 129:158–168

Gertler A, Walker A, Friesen HG (1985) Enhancement of human growth hormone-stimulated mitogenesis of Nb2 node lymphoma cells: the role of protein kinase C and calcium mobilization. Immunopharmacology 12:37–51

Gertler A, Petridou B, Kriwi GG, Djiane J (1993) Interaction of lactogenic hormones with purified recombinant extracellular domain of rabbit prolactin receptor expressed in insect cells. FEBS Lett 319:277–281

Goodman HS, Rosen JM (1990) Transcriptional analysis of the mouse beta-casein gene. Mol Endocrinol 4:1661–1670

Grunicke HH, Maly K (1993) Role of GTPases and GTPase regulatory proteins in oncogenesis (1993) Crit Rev Oncogen 4:389–402

Gu Y, Jayatilak PG, Parmer TG, Gauldie J, Fey GH, Gibori G (1992) α2-Macroglobulin expression in the mesometrial decidua and its regulation by decidual luteotropin and prolactin. Endocrinology 131:1321–1328

Happ B, Hynes NE, Groner B (1993) Ha-*ras* and v-*raf* oncogenes, but not *int–2* and c-*myc*, interfere with the lactogenic hormone dependent activation of the mammary gland specific transcription factor. Cell Growth & Differ 4:9–15

Haraguchi S, Good RA, Engelman RW, Day NK (1992) Human prolactin regulates transfected MMTV LTR-directed gene expression in a human breast-carcinoma cell line through synergistic interaction with steroid hormones. Int J Cancer 52:928–933

Harris S, McClenaghan M, Simons JP, Ali S, Clark AJ (1990) Gene expression in the mammary gland. J Reprod Fert 88:707–715

Härtig E, Nebl G, Mink S, Doppler W, Cato ACB (1993) Steroid hormone and mammary cell type-specific regulation of expression of mouse mammary tumor virus. Life Science Advances-Molecular Biology

Hayes TE, Kitchen AM, Cochran BH (1987) Inducible binding of a factor to the c-*fos* regulatory region. Proc Natl Acad Sci USA 84:1272–1276

Hennighausen L (1990) The mammary gland as a bioreactor: production of foreign proteins in milk. Protein Expression and Purification 1:3–8

Hennighausen L, Westphal C, Sankaran L, Pittius CW (1991) Regulation of expression of genes for milk proteins. In: First N, Haseltine FP (eds) Transgenic animals. Butterworth-Heinemann, Reed, New York.

Hiroka Y, Tatsumi K, Shiozawa M, Aiso S, Fukazawa T, Yasuda K, Miyia K (1991) A placental specific 5 non-coding exon of human prolactin. Mol Cell Endocrinol 75:71–80

Hitti YS, Horseman ND (1991) Structure of the gene encoding columbid annexin Icp35. Gene 103:185–192

Holtkamp W, Wuttke W, Nagel GA, Blossey HC (1988) Vergleichende Untersuchungen zum Prolaktin-, Östrogen-, Gestagen- und Androgenrezeptorgehalt menschlicher Mammakarzinome. Onkologie 11:71–76

Horlick KR, Ganjianpour M, Frost SC, Nick HS (1991) Annexin-I regulation in response to suckling and rat mammary cell differentiation. Endocrinology 128:1574–1579

Hu Z-Z, Dufau, ML (1991) Multiple and differential regulation of ovarian prolactin receptor messenger RNAs and their expression. Biochem Biophys Res Commun 181:219–225

Hynes NE, Taverna D, Harwerth IM, Ciardiello F, Salomon DS, Yamamoto T, Groner B (1990) Epidermal growth factor receptor, but not c-erbB-2, activation prevents lactogenic hormone induction of the β-casein gene in mouse mammary epithelial cells. Mol Cell Biol. 10:4027–4034

Jagoda CA, Rillema JA (1991) Temporal effect of prolactin on the activities of lactose synthetase, α-lactalbumin, and galactosyl transferase in mouse mammary gland explants. Proc Soc Exp Biol Med 197:431–434

Jahn GA, Dijane J, Houdebine L-M (1989) Inhibition of casein synthesis by progestagens in vitro: modulation in relation to concentration of hormones that synergize with prolactin. J Steroid Biochem 32:373–379

Jahn GA, Edery M, Belair L, Kelly PA, Djiane J (1991) Prolactin receptor gene expression in rat mammary gland and liver during pregnancy and lactation. Endocrinology 128:2976–2984

Jehn B, Costello E, Marti A, Keon N, Deane R, Li F, Friis RR, Burri PH, Martin F, Jaggi R (1992) Overexpression of *Mos, Ras, Src*, and *Fos* inhibits mouse mammary epithelial cell differentiation. Mol Cell Biol 12:3890–3902

Jennewein P (1992) Funktionelle Analyse des WAP Gen Promotors. Diplomarbeit, Universität Innsbruck

Jolicoeur C, Boutin J-M, Okamura H, Raguet S, Djiane J, Kelly PA (1989) Multiple regulation of prolactin receptor gene expression in rat liver. Mol Endocrinol 3:895–900

Kanai A, Nonomura N, Yoshimura M, Oka T (1993) DNA-binding proteins and their *cis*-acting sites controlling hormonal induction of a mouse β-casein-CAT fusion protein in mammary epithelial cells. Gene 26:195–201

Kelly PA, Djiane J, Postel-Vinay MC, Edery M (1991) The prolactin/growth hormone receptor family. Endocrine Rev 12:235–251

Khan KD, Shuai K, Lindwall G, Maher SE, Darnell JE Jr, Bothwell ALM (1993) Induction of the Ly-6A/E gene by interferon α/β and γ requires a DNA element to which a tyrosine-phosphorylated 91-kDa protein binds. Proc Natl Acad Sci USA 90:6806–6810

Kleinberg DL (1987) Prolactin and breast cancer. N Engl J Med 316:269–271

Kleis-SanFrancisco S, Hewetson A, Chilton BS (1993) Prolactin augments progesterone-dependent uteroglobin gene expression by modulating promoter-binding proteins. Mol Endocrinol 7:214–223

Köbberling J (1983) Akromegalie – pathophysiologische und therapeutische Grundlagen. In: Bromocriptin. Ein fachübergreifendes Therapieprinzip. Schattauer, Stuttgart, pp 135–148

Koh CY, Phillips JT (1993) Prolactin receptor expression by lymphoid tissues in normal and immunized rats. Mol Cell Endocrinol 92:R21-R25

Kurtz A, Bristol LA, Tóth BE, Lazar-Wesley E, Takács L, Kacsóh B (1993) Mammary epithelial cells of lactating rats express prolactin messenger ribonucleic acid. Biol Reprod 48:1095–1103

Larner AC, David M, Feldman GM, Igarashi K-i, Hackett RH, Webb DSA, Sweitzer SM, Petricoin EF III, Finbloom DS (1993) Tyrosine phosphorylation of DNA binding proteins by mulitple cytokines. Science 261:1730–1733

Lechleiter JD, Clapham DE (1992) Molecular mechanisms of intracellular calcium excitability in X. laevis oocytes. Cell 69:283–294

Lee CS, Oka T (1992a) Progesterone regulation of a pregnancy-specific transcription repressor to β-casein gene promoter in mouse mammary gland. Endocrinology 131:2257–2262

Lee CS, Oka T (1992b) A pregnancy-specific mammary nuclear factor involved in the repression of the mouse β-casein gene transcription by progesterone. J Biol Chem 267:5797–5801

Legraverend C, Mode A, Westin S, Ström A, Eguchi H, Zaphiropoulos PG, Gustafsson J-Å (1992) Transcriptional regulation of rat P-450 2C gene subfamily members by the sexually dimorphic pattern of growth hormone secretion. Mol Endocrinol 6:259–266

Leontic EA, Tyson JE (1977) Prolactin and fetal osmoregulation: water transport across isolated human amnion. Am J Physiol 232:R124-R127

Lesueur L, Edery M, Paly J, Kelly P, Djiane J (1990) Prolactin stimulates milk protein promoter in CHO cells cotransfected with prolactin receptor cDNA. Mol Cell Endocrinol 71:R7-R12

Lesueur L, Edery M, Ali S, Paly J, Kelly PA, Djiane J (1991) Comparison of long and short forms of the prolactin receptor on prolactin-induced milk protein gene transcription. Proc Natl Acad Sci USA 88:824–828

Leung DW, Spencer SA, Cachianes G, Hammonds G, Collins C, Henzel WJ, Barnard R, Waters MJ, Wood WI (1987) Growth hormone receptor and serum binding protein: purification, cloning and expression. Nature 330:537–543

Luck DN, Gout PW, Sutherland ER, Fox K, Huyer M, Smith M (1992) Analysis of disulphide bridge function in recombinant bovine prolactin using site-specific mutagenesis and renaturation under mild alkaline conditions – a crucial role for the central disulphide bridge in the mitogenic activity of the hormone. Protein Eng 5:559–567

Maddox PR, Jones DL, Mansel JRE (1992) Prolactin and total lactogenic hormone measured by microbioassay and immunoassay in breast cancer. Br J Cancer 65:456–460

Markoff E, Sigel MB, Lacour N, Seavey BK, Friesen HG, Lewis UJ (1988) Glycosylation selectively alters the biological activity of prolactin. Endocrinology 123:1303–1306

McKnight RA, Shamay A, Sankaran L, Wall RJ, Hennighausen L (1992) Matrix-attachment regions can impart position-independent regulation of a tissue-specific gene in transgenic mice. Proc Natl Acad Sci USA 89:6943–6947

Meyer N, Prentice DA, Fox MT, Hughes JP (1992) Prolactin-induced proliferation of the Nb2 T-lymphoma is associated with protein kinase-C-independent phosphorylation of strathmin. Endocrinology 131:1977–1984

Mieth M, Boehmer FD, Ball R, Groner B, Grosse R (1990) Transforming growth factor-β inhibits lactogenic hormone induction of β-casein expression in HC11 mouse mammary epithelial cells. Growth Factors 4:9–15

Mink S, Hartig E, Jennewein P, Doppler W, Cato ACB (1992) A mammary cell-specific enhancer in mouse mammary tumor virus DNA is composed of multiple regulatory elements including binding sites for CTF/NFI and a novel transcription factor, mammary cell-activating factor. Mol Cell Biol 12:4906–4918

Mode A, Wiersma-Larsson E, Ström A, Zaphiropoulos PG, Gustafsson, J-Å (1989) A dual role for growth hormone as a feminizing and masculinizing factor in control of sex-specific cytochrom P450 isozymes in rat liver. J Endocrinol 120:311–317

Mok E, Golovkina TV, Ross SR (1992) A mouse mammary tumor virus mammary gland enhancer confers tissue-specific but not lactation-dependent expression in transgenic mice. J Virol 66:7529–7532

Murphy LC, Tsuyuki D, Myal Y, Shiu RPC (1987) Isolation and sequencing of a cDNA clone for a prolactin-inducible protein (PIP): regulation of PIP gene expression in the human breast cancer cell line: T-47D. J Biol Chem 262:15236–15241

Murphy LC, Murphy LJ, Tsuyuki D, Duckworth ML, Shiu RPC (1988a) Cloning and characterization of a cDNA encoding a highly conserved, putative calcium binding protein, identified by an anti-prolactin receptor antiserum. J Biol Chem 263:2397–2401

Murphy PR, DiMattia GE, Friesen HG (1988b) Role of calcium in prolactin-stimulated c-*myc* gene expression and mitogenesis in Nb2 lymphoma cells. Endocrinology 122:2476–2485

Myal Y, Robinson DB, Iwasiow B, Tsuyuki D, Wong P, Shiu RPC (1991) The prolactin-inducible protein (PIP/GCDFP-15) gene – cloning, structure and regulation. Mol Cell Endocrinol 80:165–175

Nagano M, Kelly PA (1992) Absence of a putative ATP/GTP binding site in the rat prolactin receptor. Biochem Biophys Res Commun 183:610–618

Nicoll CS (1974) Physiological actions of prolactin. In: Knobil E, Saywer WH (eds) Handbook of physiology, American Physiological Society, Washington, DC, Section 7, Vol 4, Part 2, pp253-292

Nicoll CS, Herbert NJ, Russell SM (1985) Lactogenic hormones stimulate the liver to secrete a factor that acts synergistically with prolactin to promote growth of the pigeon crop-sac mucosal epithelium in vivo. Endocrinology 116:1449–1453

Nishizuka Y (1992) Intracellular signaling by hydrolysis of phospholipids and activation of protein kinase C. Science 258:607–614

O'Neal KD, Schwarz LA, Yu-Lee L-y (1991) Prolactin receptor gene expression in lymphoid cells. Mol Cell Endocrinol 82:127–135

O'Neal KD, Montgomery DW, Truong TM, Yu-Lee L-y (1992) Prolactin gene expression in human thymocytes. Mol Cell Endocrinol 87:R19-R23

Ono M, Oka T (1980) The differential actions of cortisol on the accumulation of α-lactalbumin and casein in midpregnant mouse mammary gland in culture. Cell 19:473–480

Ormandy CJ, Sutherland RL (1993) Mechanisms of prolactin receptor regulation in mammary gland. Mol Cell Endocrinol 91:C1-C6

Ormandy CJ, Clarke CL, Kelly PA, Sutherland RL (1992) Androgen regulation of prolactin-receptor gene expression in MCF-7 and MDA-MB-453 human breast cancer cells. Int J Cancer 50:777–782

Ormandy CJ, Lee CSL, Kelly PA, Sutherland RL (1993) Regulation of prolactin receptor expression by the tumour promoting phorbol ester 12-O-tetradecanoylphorbol-13-acetate in human breast cancer cells. J Cell Biochem 52:47–56

Parry G, Stubbs J, Bissell MJ, Schmidhauser C, Spicer AP, Gendler SJ (1992) Studies of Muc-1 mucin expression and polarity in the mouse mammary gland demonstrate developmental regulation of Muc-1 glycosylation and establish the hormonal basis for mRNA expression. J Cell Science 101:191–199

Pearse RN, Feinman R, Shuai K, Darnell JE Jr, Ravetch JV (1993) Interferon γ-induced transcription of the high-affinity Fc receptor for IgG Requires assembly of a complex that includes the 91-kDa subunit of transcription factor ISGF3. Proc Natl Acad Sci USA 90:4314–4318

Pellegrini I, Lebrun JJ, Ali S, Kelly PA (1992) Expression of prolactin and its receptor in human lymphoid cells. Mol Endocrinol 6:1023–1031

Pérez-Villamil B, Bordiú E, Puente-Cueva M (1992) Involvement of physiological prolactin levels in growth and prolactin receptor content of prostate glands and testes in developing male rats. J Endocrinol 132:449–459

Peters BJ, Rillema JA (1992) Effect of prolactin on 2-deoxyglucose uptake in mouse mammary gland explants. Am J Physiol 262:E627-E630

Pierre S, Jolivet G, Devinoy E, Théron MC, Maliénou-N'Gassa R, Puissant C, Houdebine LM (1992) A distal region enhances the prolactin induced promoter activity of the rabbit αs1-casein gene. Mol Cell Endocrinol 87:147–156

Pittius CW, Sankaran L, Topper YJ, Hennighausen L (1988) Comparision of the regulation of the whey acidic protein gene with that of a hybrid gene containing the whey acidic protein gene promoter in transgenic mice. Mol Endocrinol 2:1027–1032

Posner BI, Kelly PA, Friesen HG (1974) Induction of a lactogenic receptor in rat liver: influence of estrogen and the pituitary. Proc Natl Acad Sci USA 71:2407–2410

Postel-Vinay MC, Belair L, Kayser C, Kelly PA, Djiane J (1991) Identification of prolactin and growth hormone binding proteins in rabbit milk. Proc Natl Acad Sci USA 88:6687–6690

Prins GS (1987) Prolactin influence on cytosol an nuclear androgen receptors in the ventral, dorsal, and lateral lobes of the rat prostate. Endocrinology 120:1457–1464

Prosser CG, Sankaran L, Hennighausen L, Topper YJ (1987) Comparison of the roles of insulin and insulin-like growth factor i in casein gene expression and in the development of α-lactalbumin and glucose transport activities in the mouse mammary epithelial cell. Endocrinology 120:1411–1416

Puissant C, Houdebine LM (1991) Cortisol induces rapid accumulation of whey acid protein messenger RNA but not of αSl and β-casein mRNA in rabbit mammary explants. Cell Biol Int Rep 15:121–129

Randall GW, Daniel JC, Chilton BS (1991) Prolactin enhances uteroglobin gene expression by uteri of immature rabbits. J Reprod Fertil 91:249–257

Ray DB, Jansen RW, Horst IA, Mills NC, Kowal J (1986) A complex noncoordinate regulation of α-lactalbumin and 25k β-casein by corticosterone, prolactin, and insulin in long term cultures of normal rat mammary cells. Endocrinology 118:393–407

Reichmann E, Ball RK, Groner B, Friis RR (1989) New mammary epithelial and fibroblastic cell clones in coculture form structures competent to differentiate functionally. J Cell Biol 108:1127–1138

Rillema JA, Etindi RN, Ofenstein JP, Waters SB (1988) Mechanisms of prolactin action. In: Knobil E, Neill J(eds) The physiology of reproduction. Raven, New York, pp2217-2234

Rillema JA, Tarrant TM, Linebaugh BE (1989) Studies on the mechanism by which prolactin regulates protein, RNA, and DNA synthesis in NB2 node lymphoma cells. Biochim Biophys Acta 1014:78–82

Rillema JA, Campbell GS, Lawson DM, Carter-Su C (1992) Evidence for a rapid stimulation of tyrosine kinase activity by prolactin in Nb2 rat lymphoma cells. Endocrinology 131:973–975

Robinson SD, Roberts AB, Daniel CW (1993) TGF β suppresses casein synthesis in mouse mammary explants and may play a role in controlling milk levels during pregnancy. J Cell Biol 120:245–251

Rosen, JM, Rodgers JR, Couch CH, Bisbee CA, David-Inouye Y, Campbell SM, Yu-Lee K-Y (1986) Multihormonal regulation of milk protein gene expression. Ann NY Acad Sci 478:63–76

Rozakis-Adcock M, Kelly PA (1991) Mutational analysis of the ligand-binding domain of the prolactin receptor. J Biol Chem 266:16472–16477

Ruff-Jamison, Chen K, Cohen S (1993) Induction by EGF and interferon-γ of tyrosine phosphorylated DNA binding proteins in mouse liver nuclei. Science 261: 1733–1736

Rui H, Djeu JY, Evans GA, Kelly PA, Farrar WL (1992) Prolactin receptor triggering – evidence for rapid tyrosine kinase activation. J Biol Chem 267:24076–24081

Sadowski HB, Shuai K, Darnell JE Jr, Gilman MZ (1993) A common nuclear signal transduction pathway activated by growth factor and cytokine receptors. Science 261:1739–1744

Sakai S, Yamamoto K, Aihara M, Suzuki M, Nagasawa H (1990) Prolactin and progesterone receptors in pregnancy-dependent mammary tumors in GR/A mice. Proc Soc Exp Biol Med 195:375–378

Sankaran L, Topper YL (1988) Progesterone and prolactin are both required for suppression of the induction of rat α-lactalbumin activity. Biochem Biophys Res Commun 155:1038–1045

Scammell JG, Luck DN, Valentine DL, Smith M (1992) Epitope mapping of monoclonal antibodies to bovine prolactin. Am J Physiol 263:E520-E525

Schindler C, Shuai K, Prezioso VR, Darnell JE Jr (1992) Interferon-dependent tyrosine phosphorylation of a latent cytoplasmic transcription factor. Science 257:809–813

Schmidhauser C, Bissell MJ, Myers CA, Casperson GF (1990) Extracellular matrix and hormones transcriptionally regulate bovine β-casein 5' sequences in stably transfected mouse mammary cells. Proc Natl Acad Sci USA 87:9118–9122

Schmidhauser C, Casperson GF, Myers CA, Sanzo KT, Bolten S, Bissell MJ (1992) A novel transcriptional enhancer is involved in the prolactin and extracellular matrix-dependent regulation of β-casein gene expression. Mol Biol Cell 3:699–709

Schmitt-Ney M, Doppler W, Ball RK, Groner B (1991) β-Casein gene promoter activity is regulated by the hormone-mediated relief of transcriptional repression and a mammary-gland-specific nuclear factor. Mol Cell Biol 11:3745–3755

Schmitt-Ney M, Happ B, Ball RK, Groner B (1992a) Developmental and environmental regulation of a mammary gland-specific nuclear factor essential for transcription of the gene encoding β-casein. Proc Natl Acad Sci USA 89:3130–3134

Schmitt-Ney M, Happ B, Hofer P, Hynes NE, Groner B (1992b) Mammary gland-specific nuclear factor activity is positively regulated by lactogenic hormones and negatively by milk stasis. Mol Endocrinol 6:1988–1997

Schoenenberger C-A, Zuk A, Groner B, Jones W, Andres A-C (1990) Induction of the endogenous whey acidic protein (Wap) Gene and a *Wap-myc* hybrid gene in primary murine mammary organoids. Development Biol 139:327–337

Schwarz LA, Stevens AM, Hrachovy JA, Yu-Lee L-y (1992) Interferon regulatory factor-1 is inducible by prolactin, interleukin-2 and concanavalin A in T cells. Mol Cell Endocrinol 86:103–110

Scott P, Kessler MA, Schuler LA (1992) Molecular cloning of the bovine prolactin receptor and distribution of prolactin and growth hormone receptor transcripts in fetal and utero-placental tissues. Mol Cell Endocrinol 89:47–58

Shirota M, Banville D, Ali S, Jolicoeur C, Boutin J-M, Edery M, Dijane J, Kelly PA (1990) Expression of two forms of prolactin receptor in rat ovary and liver. Mol Endocrinol 4:1136–1143

Shiu RPC (1979) Prolactin receptors in human breast cancer cells in long-term tissue culture. Cancer Res 39:4381–4386

Shiu RPC, Iwasiow BM (1985) Prolactin-inducible proteins in human breast cancer cells. J Biol Chem 260:11307–11313

Shuai K, Stark GR, Kerr IM, Darnell JE Jr (1993) A single phosphotyrosine residue of Stat91 required for gene activation by interferon-γ. Science 261:1744–1746

Silvennoinen O, Schnidler C, Schlessinger J, Levy DE (1993) *Ras*-independent growth factor signaling by transcription factor tyrosine phosphorylation. Science 261:1736–1739

Sinha YN, Depaolo LV, Haro LS, Singh RNP, Jacobsen BP, Scott KE, Lewis UJ (1991) Isolation and biochemical properties of four forms of glycosylated porcine prolactin. Mol Cell Endocrinol 80:203–213

Skwarlo-Sonta K (1992) Prolactin as an immunoregulatory hormone in mammals and birds. Immunol Lett 33:105–122

Soares MJ, Faria TN, Roby KF, Deb S (1991) Pregnancy and the prolactin family of hormones: coordination of anterior pituitary, uterine, and placental expression. Endocrine Rev 12:402–423

Stahl N, Yancopoulos GD (1993) The alphas, betas and kinases of cytokine receptor complexes. Cell 74:587–590

Staindl B, Berger P, Kofler R, Wick G (1987) Monoclonal antibodies against human, bovine and rat prolactin: epitope mapping of human prolactin and development of a two-site immunoradiometric assay. J Endocrinol 114:311–318

Steinmetz RW, Grant AL, Malven PV (1993) Transcription of prolactin gene in milk secretory cells of the rat mammary gland. J Endocrinol 136:271–271

Stevens AM, Yu-Lee L-y (1992) The transcription factor interferon regulatory factor-1 is expressed during both early G1 and the G1/S transition in the prolactin-induced lymphocyte cell cycle. Mol Endocrinol 6:2236–2243

Strange R, Li F, Friis RR, Reichmann E, Haenni B, Burri PH (1991) Mammary epithelial differentiation in vitro – minimum requirements for a functional response to hormonal stimulation. Cell Growth Differ 2:549–559

Stricker, P, Grueter, R (1928) Action du lobe anterieur de lhypophse sur la montée laiteuse. Compt Rend Soc Biol 99:1978–1980

Taketani Y, Oka T (1983) Tumor promoter 12-o-tetradecanoylphorbol 13-acetate, like epidermal growth factor, stimulates cell proliferation and inhibits differentiation of mouse mammary epithelial cells in culture. Proc Natl Acad Sci USA 80:1646–1649

Tata JR, Kawahara A, Baker BS (1991) Prolactin inhibits both thyroid hormone-induced morphogenesis and cell death in cultured amphibian larval tissues. Dev Biol 146:72–80

Taverna D, Groner B, Hynes NE (1991) Epidermal growth factor receptor, platelet-derived growth factor receptor, and c-erbB-2 receptor activation all promote growth but have distinctive effects upon mouse mammary epithelial cell differentiation. Cell Growth Differ 2:145–154

Thordarson G, Fielder P, Chul L, Yun KH, Robleto D, Ogren L, Talamantes F (1992) Mammary gland differentiation in hypophysectomized, pregnant mice treated with corticosterone and thyroxine. Biol Reprod 47:676–682

Thoreau E, Petridou B, Kelly PA, Djiane J, Mornon JP (1991) Structural symmetry of the extracellular domain of the cytokine/growth hormone/prolactin receptor family and interferon receptors revealed by hydrophobic cluster analysis. FEBS Lett 282:26–31

Tiong TS, Herington AC (1992) Ontogeny of messenger RNA for the rat growth hormone receptor and serum binding protein. Mol Cell Endocrinol 83:133–141

Too CKL, Shiu RPC, Friesen HG (1990) Cross-linking of G-proteins to the prolactin receptor in rat Nb2 lymphoma cells. Biochem Biophys Res Commun 173:48–52

Topper YJ, Freeman CS (1980) Multiple hormone interactions in the developmental biology of the mammary gland. Physiol Rev 60:1049–1106

Turner MD, Rennison ME, Handel SE, Wilde CJ, Burgoyne RD (1992) Proteins are secreted by both constitutive and regulated secretory pathways in lactating mouse mammary epithelial cells. Endocrinology 117:269–278

Ullrich A, Schlessinger J (1990) Signal transduction by receptors with tyrosine kinase activity. Cell 61:203–212

Ultsch M, Devos AM, Kossiakoff AA (1991) Crystals of the complex between human growth hormone and the extracellular domain of its receptor. J Mol Biol 222:865–868

Villalba M, Zabala MT, Martinez-Serrano A, de la Colina R, Satrústegui J, Garcia-Ruiz JP (1991) Prolactin increases cytosolic free calcium concentration in hepatocytes of lactating rats. Endocrinology 129:2857–2861

Vizoso F, Sanchez LM, Diezitza I, Lamelas ML, Lopezotin C (1992) Factors affecting protein composition of breast secretions from nonlactating women. Breast Cancer Res Treat 23:251–258

Vonderhaar BK, Biswas R (1987) Prolactin effects and regulation of its receptors in human mammary tumor cells. In: Medina D, Kidwell W, Heppner G, Anderson E (eds) Cellular and molecular biology of mammary cancer. Plenum Press, New York, pp 205-219

Voss JW, Rosenfeld MG (1992) Anterior pituitary development: short tales form dwarf mice. Cell 70:527–530

Wakao H, Schmitt-Ney M, Groner B (1992) Mammary gland-specific nuclear factor is present in lactating rodent and bovine mammary tissue and composed of a single polypeptide of 89 kDa. J Biol Chem 267:16365–16370

Waters SB, Rillema JA (1989) Role of protein kinase C in the prolactin-induced responses in mouse mammary gland explants. Mol Cell Endocrinol 63:159–166

Wegenka UM, Buschmann J, Lütticken C, Heinrich PC, Horn F (1993) Acute-phase response factor, a nuclear factor binding to acute-phase response elements, is rapidly activated by interleukin-6 at the posttranslational level. Mol Cell Biol 13: 276–288

Welte T, Philipp S, Cairns C, Gustafsson J-Å, Doppler W (1993) Glucocorticoid receptor binding sites in the promoter region of milk protein genes. J Steroid Biochem (in press).

Wilks AF, Harpur AG, Kurban RR, Ralph SJ, Zürcher G, Ziemiecki A (1991) Two novel protein-tyrosine kinases, each with a second phosphotransferase-related catalytic domain, define a new class of protein kinase. Mol Cell Biol 11:2057–2065

Wilson KC, Finbloom DS (1992) Interferon γ rapidly induces in human monocytes a DNA-binding factor that recognizes the gamma response region within the promoter of the gene for the high-affinity Fc γ receptor. Proc Natl Acad Sci USA 89:11964–11968

Wolff J, Wong C, Cheng H, Poyet P, Butel JS, Rosen JR (1992) Differential effects of the simian virus 40 early genes on mammary epithelial cell growth, morphology, and gene expression. Exp Cell Res 202:67–76

Xu Y-H, Horseman ND (1992) Nuclear proteins and prolactin-induced *Annexin I*c_{p35} gene transcription. Mol Endocrinol 6:375–383

Yoon J-B, Berry SA, Seelig S, Towle HC (1990) An inducible nuclear factor binds to a growth hormone-regulated gene. J Biol Chem 265:19947–19954

Yoshimura M, Oka T (1990a) Transfection of β-casein chimeric gene and hormonal induction of its expression in primary murine mammary epithelial cells. Proc Natl Acad Sci USA 87:3670–3674

Yoshimura M, Oka T (1990b) Hormonal induction of β-casein gene expression: requirement of ongoing protein synthesis for transcription. Endocrinology 126:427–433

Yu-Lee L-y (1990) Prolactin stimulates transcription of growth related genes in Nb2 T lymphoma cells. Mol Cell Endocrinol 68:21–28

Yu-Lee L-y, Hrachovy JA, Stevens AM, Schwarz LA (1990) Interferon regulatory factor-1 is an immediate-early gene under transcriptional regulation by prolactin in Nb2 T cells. Mol Cell Biol 10:3087–3094

Zabala MT, Garcia-Ruiz JP (1989) Regulation of expression of the messenger ribonucleic acid encoding the cytosolic form of phosphoenolpyruvate carboxykinase in liver and small intestine of lactating rats. Endocrinology 125:2587–2593

Zeigler ME, Wicha MS (1992) Posttranscriptional regulation of α-casein mRNA accumulation by laminin. Exp Cell Res 200:481–489

Zhang R, Buczko E, Tsai-Morris C-H, Hu Z-Z, Dufau M (1990) Isolation and characterization of two novel rat ovarian lactogen receptor cDNA species. Biochem Biophys Res Commun 168:415–422

Editor-in-charge: Prof. H. Grunicke and Prof. H. Schweiger

Subject Index

Springer-Verlag and the Environment

We at Springer-Verlag firmly believe that an international science publisher has a special obligation to the environment, and our corporate policies consistently reflect this conviction.

We also expect our business partners – paper mills, printers, packaging manufacturers, etc. – to commit themselves to using environmentally friendly materials and production processes.

The paper in this book is made from low- or no-chlorine pulp and is acid free, in conformance with international standards for paper permanency.

Printing: Mercedesdruck, Berlin
Binding: Buchbinderei Lüderitz & Bauer, Berlin